T0247607

Memory and Movies

Memory and Movies

What Films Can Teach Us about Memory

John Seamon

The MIT Press
Cambridge, Massachusetts
London, England

This book was set in Stone Sans and Stone Serif by the MIT Press.

Library of Congress Cataloging-in-Publication Data

Seamon, John G., 1943–
Memory and movies : what films can teach us about memory / John Seamon.
pages cm
Includes bibliographical references and index.
ISBN 978-0-262-02971-1 (hardcover : alk. paper)
ISBN 978-0-262-55329-2 (paperback)
1. Memory. 2. Cognition. 3. Motion pictures. I. Title.
BF385.S36 2015
153.1'2—dc23
2015009377

148364161

For my wife Diane … always.

Contents

Preface

The origin of this book can be traced to the release of the film *Memento* in 2000. Wondering whether popular films could enrich learning, I tried an experiment in my memory course at Wesleyan University. After describing different memory pathologies, I showed this movie to my students and asked them to write an essay detailing what *Memento* got right about memory and what it got wrong. I had no idea how these people would respond, especially because many of them had already seen the film. But my fears were quickly allayed when they responded enthusiastically to the exercise and said that they saw the film in a new light. No longer was *Memento* seen as a bizarre revenge story, cleverly told in backward chronological order—now it became a vehicle for demonstrating their new understanding of memory. Their enthusiasm led me to develop a new course—*Memory in the Movies*—and, ultimately, to writing this book, both focusing on understanding memory through popular films.

Memory and Movies is not a college text. I wrote it for a general audience with no background in psychology. It is for anyone who loves watching movies and seeks a basic understanding of memory that is grounded in scientific knowledge. Over eight chapters, I sketch a "big picture" view of memory, covering its major phenomena in a nontechnical manner and illustrating them with a wide variety of films—including many movies that readers may have already seen. But as my students experienced in viewing *Memento*, examining these films with an educated eye sheds new light on memory function and its dysfunction, promoting a greater understanding of this fundamental aspect of mind.

This book has my name on the cover, but others helped me to shape it. I offer my sincerest appreciation to each of the following: the many Wesleyan students who responded enthusiastically to my memory and film course and convinced me that movies could enhance education; my colleague and friend, Charles Sanislow, who tirelessly read chapter drafts and made each one better by sprinkling encouragement with his astute criticism; my editor, Philip Laughlin, who, when I submitted an initial inquiry, asked for the entire manuscript and proceeded to make me an MIT Press author; the people at MIT Press, including Christopher Eyer, Judith Feldmann, Susan Clark, and Molly Seamans, who turned my manuscript into a book; and finally, Wesleyan University, for providing a wonderfully supportive environment for over forty years and giving me the freedom to keep reinventing my courses and myself. I am indebted to all. Thank you.

John Seamon
December 9, 2014

1 Memory Processes and Memory Films

Setting the Scene

Our memory is a collection of systems that serve different functions but work together in a unified fashion. More than a record of past experiences, memory enables us to imagine and plan future action, similar to the role of stories in movies. By observing the experiences of others, we can imagine our own responses to comparable situations in the future. Films tell us stories about people, and they can enhance our understanding of memory by illustrating the diverse ways in which the present is influenced by the past.

Sometimes Life Imitates Art

In Peter Segal's romantic comedy, *50 First Dates*, Drew Barrymore plays Lucy, a young woman experiencing a bizarre form of amnesia. Lucy, we are told, suffered a brain-altering concussion from a car crash, stopping her memory on that date. She remembers what happened before the crash, and she remembers each new day's events as they unfold, but those new memories are quickly forgotten each night when she goes to sleep. She awakens each day to visual reminders that begin: "Good morning,

Lucy. Here are some things you missed this year." Lucy can no longer remember one day from the next.

A serious brain concussion can produce major memory impairment, but not normally in the way it is depicted in this film. Brain damage can lead to profound problems in making those long-lasting memories that give continuity to our daily experiences. But Lucy's memory could not function normally during the day, only to be lost at night. Instead, if her brain damage were severe enough to impair those processes vital for making new memories, she would have a hard time retaining new experiences for more than a few minutes. Her daily living would become nearly totally dependent on the use of external notes, diaries, and recordings—weak substitutes for losing her internally functioning memory.

Yet, one year after the release of *50 First Dates*, a woman, identified by the initials FL, was involved in an automobile accident and showed the same memory problems as Lucy.[1] FL remembered her past prior to the accident, and she remembered events as they occurred each day after the accident, but those new memories disappeared each night during sleep. FL puzzled researchers because her neurological exam was normal, as were her brain scans and psychiatric evaluation. Although FL showed no evidence of brain dysfunction, the researchers did not think that she was feigning amnesia. FL believed that she had the memory impairment that was depicted in *50 First Dates*, a film that she had seen before her accident, starring her favorite actress. The researchers suspected that FL was experiencing *functional amnesia*, amnesia with a psychological origin, rather than *organic amnesia*, amnesia with a biological basis.[2] Their hunch was bolstered when they found that FL could remember pictures that she was shown the same day as testing as well as pictures that she thought had also been shown the same day, but were actually shown from a preceding day.

The take-home message from the story of FL is not that movies are an inherently misleading way to learn about memory. Sometimes movies get memory right. But in using films to learn about memory, we need to be mindful of this difference and let our viewing be guided by scientific knowledge, rather than commonsensical or nonsensical depictions of memory in film. This book provides that knowledge, starting with an overview of memory.

What Is Memory?

Psychologist Endel Tulving once described memory as "a gift of nature, the ability of living organisms to retain and utilize acquired information or knowledge."[3] This appealing definition still works, but it does not speak to the complexity of memory as we now understand it. For example, all of us have heard other people describe their "memory" as bad or good, as though memory were a single aspect of mind that functioned either poorly or well. Researchers, in fact, used to think of memory this way, but our concept of memory has evolved, and today it is richer and far more complex.

A Collection of Memory Systems

Over recent decades, Tulving and others have shown that what we call memory is best thought of as a diverse collection of independent systems designed to serve different purposes. These multiple memory systems, involving different brain structures and networks, include working memory, episodic memory, semantic memory, and procedural memory.

Working memory is sometimes called the mind's workbench because it enables us to maintain a small number of thoughts in conscious awareness. We can use it to think verbally—as in remembering a phone number by mentally rehearsing the digits—or visually—as in imagining how our house would look if painted a different color. As soon as our attention is drawn elsewhere, prior thoughts are quickly forgotten, replaced by new ones with each attention shift. Working memory is temporary, ideal for short-term retention. The other memory systems are more permanent, allowing us to retain a vast amount of information for an indefinite period of time.

Episodic memory refers to the experiences that make up our life. It represents our conscious recollection of the past, the type of memory we use when we mentally travel backward in time, as in recalling a favorite book, a joyous summer vacation, or our first kiss. In talking about our recollections, we are almost always referring to our episodic long-term memory, the memory system that will be the primary focus of this book.

Semantic memory represents the factual knowledge that we have acquired, such as the meaning of the word *democracy*, the fact that 2 + 2 = 4, and the fact that fire engines are painted yellow for better visibility at night. Whereas episodic memory is often specific to a particular time and place, as in your memory of your first kiss, semantic memory is more generic and not linked to specific events, as in your understanding of the meaning of a kiss.

Finally, *procedural memory* represents our knowledge of how to perform a complex activity. We demonstrate this type of long-term memory by performing a skilled action that was acquired slowly through trial and error learning. Whereas episodic or semantic memory may be declared, procedural memory can only be demonstrated. Whether riding a bicycle or driving a car, we demonstrate procedural memory by performance, not verbal recall. Merely explaining the basic mechanics of swimming to a nonswimmer, for instance, would be useless for the would-be swimmer. Procedural memories require extensive practice. Once acquired, they can last a lifetime.

We know that these memory systems can function independently because people with different forms of brain damage can have an impairment in one or another of these systems, while the others are unaffected. For example, a person might lose the ability to produce new episodic memories following a concussion, yet still be able to acquire new procedural skills. Although any of these memory systems might become impaired, episodic memory is our most fragile system; it develops slowly during childhood and is the first to be impaired in old age. A person with a dysfunctional episodic memory may demonstrate *retrograde amnesia*, difficulty in recalling old memories, or *anterograde amnesia*, problems in forming new memories.

A Seamless Operation of Systems

Describing the memory systems separately highlights their important differences, but they normally operate in concert with little thought or direction on our part. For example, consider the demands we place on these systems simply in taking a friend to the movies. In planning a night out, you access the movie app on your computer—*procedural memory*; read the reviews of films that are currently playing—*semantic memory*; and select one film because you enjoyed a previous film by that actor—*episodic memory*. You call your friend on your cell phone—*procedural memory*; recognize her voice when she answers—*episodic memory*; discuss the various film options—*working and semantic memory*; and, later, drive to the theater together—*procedural memory*.

Occasionally, one or another of these systems might fail us, as when we forget the location of the car keys—*episodic memory*—or the easiest route to a friend's home—*procedural memory*. But, remarkably, we use these systems repeatedly, and their seamless operation allows us to take them for granted. Memory clearly involves more than remembering, and this example provides a strong hint about its indispensable function.

Memory's Essential Function

The scientific study of memory began in 1885 when German psychologist Hermann Ebbinghaus examined various factors that influenced his ability to remember lists of unfamiliar items.[4] Researchers since then have thoroughly explored how many words, pictures, and stories people are able to remember and how long they are able to remember them. For example, my students and I recently tested a man who spent eight years memorizing all 10,565 lines of Milton's epic poem *Paradise Lost*.[5] When we gave him two lines from various sections of the poem, he recited the lines that followed with considerable accuracy. This work demonstrates the flexibility of memory and what can be accomplished with enormous dedication and practice. However, it does not speak to memory's function in everyday life. Until recently, researchers have paid little attention to the purpose of memory, other than to view it as a useful evolutionary adaptation that enables us to retain a personal record of the past. This record-keeping view still holds, but researchers now see memory's function as much more than just a way of remembering the past.

Curiously, one factor that spurred an interest in the function of everyday memory was work done on *memory illusions*.

These illusions, says psychologist Henry Roediger, are instances in which our recollections of past events differ significantly from the actual occurrence of those events.[6] Consider, for example, the following list of words—read the words once, before continuing with the text—I will explain its purpose shortly:

Bed, Rest, Awake, Tired, Dream, Snooze, Blanket, Doze, Slumber, Snore, Nap, Yawn

Understanding how memory illusions can fool us is important for understanding how we can misremember the past.[7] We will see in a later chapter that when a false childhood event—such as getting lost in a mall—is described in the context of real childhood memories by someone we trust—such as a parent or older sibling—and when we repeatedly imagine this event over time, many of us can come to believe that the imagined event actually occurred.[8] False memories can be induced by imagination even for unusual or bizarre actions, as my students and I discovered when people remembered proposing marriage to a Pepsi machine, something they had merely imagined doing on an earlier campus walk.[9] Our episodic memories are usually dependable guides to our past, but such findings indicate that they are not always so. Sometimes our memories are illusory.

As an example, without looking back at the word list in the previous paragraph, take a moment and consider each of these words—*Bed, Train,* and *Sleep.* Were any of these words on the earlier list? If you are like many people, you will confidently remember two words as part of the list. However, if you now look back, you will see that only one recollection is correct; the other is a memory illusion because it was not part of the original list. Surprisingly, this illusion occurs even if you are told about it before studying the list.[10]

The ease of demonstrating false memory raises the following question: Why would a system that generates errors in remembering have ever evolved? Computer memory, as shown by the latest laptops and iPads, differs from human memory in that computers can retain vast amounts of information—whether phone numbers, store locations, or books by a favorite author—perfectly and indefinitely, and this information is almost instantly accessible. Human memory never evolved in this fashion. Our episodic memory normally works well enough for us to function effectively—otherwise we would be in real trouble as a species—but it is subject to various errors, biases, and distortions. Psychologist Frederic Bartlett, for example, observed long ago that when we read stories, our memory of those stories is never perfect. Typically, we are able to recall the gist of a story well, but parts of the story might be vague or incomplete in our mind, and at times we unknowingly fill in those memory gaps by making inferences on the basis of what we reason must have happened.[11]

Because false recollections sometimes occur, human memory differs from computer memory, and this finding indicates that our memory is more *reconstructive* than *reproductive* in nature. To understand this important distinction, consider psychologist Ulric Neisser's apt analogy. He said that in remembering the past we function like paleontologists, who—after unearthing bits and fragments of a dinosaur's remains—reconstruct the dinosaur in its entirety from those recovered fragments.[12] This analogy is important for thinking correctly about memory. It says that our episodic memory is not a collection of snapshots of past events filed away in the brain.

Instead, when we perceive an event, we form an interpretation of that experience, and this process of interpretation is influenced by a host of factors. Memories, says psychologist Marcia

Johnson, "are attributions that we make about our mental experiences based on their subjective qualities, our prior knowledge and beliefs, our motives and goals, and the social context."[13] This is why two people can remember the same event differently: their interpretations of that event differed. Our perception of an event is not the same as taking a picture of it. Episodic memory represents a record of our perceptual interpretations, not literal copies of experienced events.

This nonobvious point is easily demonstrated. Ask a friend to look out of a window that shows a portion of a building or large tree. After fifteen seconds, have your friend look away and draw the scene from memory, saying that it is a test of memory, not artistic ability. Invariably, your friend will draw more of the building or tree than was seen through the window, a phenomenon that psychologist Helene Intraub calls the *boundary extension illusion*.[14] We all know that buildings and trees do not exist in the world incomplete; your friend's implicit knowledge that buildings and trees continue beyond the frame of a window serves to bias the drawing. The memory drawing is an interpretation of what was seen, not a literal copy of what was seen. Our memory is guided by our perceptual interpretations.

Yet, if our recollections are reconstructed based on what we know, and memory evolved so that we can acquire and utilize knowledge, why does it function imperfectly? According to psychologist Daniel Schacter, the answer may be deceptively simple. A reconstructive memory system, he says, provides for an important trade-off. It occasionally produces false recollections because it does not store perfect copies of the past, but it offers an invaluable mental ability to compensate for the loss of perfection, namely, the flexibility of allowing our interpretations of past events to be recombined in novel ways to imagine and plan the future.[15] This critically important survival adaptation, essential for imagining and planning, may be episodic memory's primary function. We can remember *Paradise Lost* verbatim if we want to, but normally we just need to understand what its author said to put that information to future use. At the cost of occasionally misremembering, our reconstructive episodic memory provides us the mental flexibility needed for traveling backward or forward in time.

Once Upon a Time ...

Long before most people could read or write, there was an established oral tradition of passing down knowledge and values in stories. Epics such as the Babylonian *Gilgamesh*, the Greek *Iliad* and *Odyssey*, and the Anglo-Saxon *Beowulf* were orally transmitted for ages before they were eventually committed to text.[16] Good stories have always been entertaining and informative, and we may harbor a need from our long-distant past to hear and tell them.[17] Neuroscientist Antonio Damasio believes that storytelling, which occurs in all societies and cultures, is something our brains naturally do.[18] This need, expressed early in life when children ask for a story before bed, remains strong over our lifespan as we spend countless hours absorbing stories by talking with friends, reading novels, and watching movies. Good stories involve various challenges, and remembering them allows us to imagine how we might navigate similar life challenges in the future. These stories are not just for entertainment, says psychologist Keith Oatley; they function as the mind's flight simulator. The brain networks that are activated by the sights, sounds, and movements of real life are also activated by reading a story.[19] Even when we go to sleep, notes author Jonathan Gottschall, our minds stay awake, telling stories.[20]

Because we are adept at remembering stories, I adopted a story format in this book. Each chapter tells a story about one major aspect of memory. Within each chapter, films provide different stories to illustrate phenomena tied to a chapter's theme. Together, the chapters and films tell a coherent story about memory and how we have come to understand it. The chapters can be read in sequence, but this order is not essential. Each chapter was designed to tell its own tale, and each can stand on its own. The chapters deal primarily with episodic memory because it is our most thoroughly studied system and the one we are most aware of by its occasional lapses. It is also the memory system most directly represented in film.

To help tell these stories about memory, I have selected forty films that not only provide clear illustrations of memory phenomena, but also, hopefully, are worthwhile to watch on their own. Virtually all of these films were made after 1980 with the majority after 2000, ensuring their ready availability. Many other relevant films are briefly noted. These movies cover a variety of film genres including comedy, drama, suspense, fantasy, science fiction, animation, and documentary. Many are American-made, but there are also notable international films involving people from different cultures, all with subtitles available.

Using Films to Tell Stories about Memory

Popular films are designed for entertainment, but they can do a great deal more. Stories in movies can inform us, and writers know that stories can change how we think by letting us glimpse life through a character's eyes.[21] How does a film correspond to life? Think back to the boundary extension illusion where we saw how our implicit knowledge tells us that there is more to a real-life scene than we can glimpse through the frame of a window. Watching a movie on a screen is like looking at life through a window—a window opening onto the world. Much more exists beyond the screen borders as the film transports us out of our everyday lives into the lives of its characters by the creative blending of light, color, and sound.[22]

Psychologists have studied the cognitive and neural processes involved in our perception of film—how it is, say Jeffrey Zacks and Joseph Magliano, that "a series of still pictures are projected quickly on a screen, accompanied by a stream of sound—and a viewer has an experience that can be as engaging, emotionally affecting, and memorable as many experiences in real life."[23] Much work has focused on understanding the filmmaker's techniques that give rise to this experience.[24] But this book is not about filmmaking; it is about memory. In explaining how memory functions, I use films to illustrate well-studied memory phenomena. Movies can inspire and inform us by opening our eyes, allowing us to see things differently. For the films assembled in this book, I want you to see how movies can inform us about memory. We begin, as always in the movies, with previews—sketches of the stories about memory that I will tell.

The Coming Attractions

Chapter 2: Movies and the Mind's Workbench

Working memory has been called the workbench of the conscious mind. Filmmakers use it to make movies and viewers need it to understand them. Some films require working memory to connect diverse story lines into a coherent narrative, whereas other films actively resist any coherent integration, existing only moment-to-moment in conscious awareness. Still other films enlist the creative aspects of working memory, enabling our imagination to soar.

Films Involving Working Memory

- *Fantasia*, animation by Walt Disney Studios (1940).
- *Life of Pi*, directed by Ang Lee, with Suraj Sharma, Irrfan Khan, and Ayush Tandon (2012).
- *Pulp Fiction*, directed by Quentin Tarantino, with Samuel L. Jackson and John Travolta (1994).
- *Mulholland Drive*, directed by David Lynch, with Naomi Watts, Laura Harring, and Justin Theroux (2001).
- *Avatar*, directed by James Cameron, with Sam Worthington, Zoë Saldana, and Sigourney Weaver (2009).

Chapter 3: Making Memories That Last

Memory involves acquiring and using information and knowledge, processes that depend on attention. Some methods of paying attention are especially effective for producing long-lasting memories; others are not. While virtually anyone can accomplish surprising feats of memory with sufficient practice and diligence, having an encyclopedic store of knowledge may be pointless unless it serves a useful everyday function.

Films Involving Long-Lasting Memories

- *After Life*, drama directed by Hirokazu Kore-eda, with Arata, Erika Oda, and Taketoshi Naito (1998, Japanese with English subtitles).
- *Cast Away*, directed by Robert Zemeckis, with Tom Hanks and Helen Hunt (2000).
- *Groundhog Day*, directed by Harold Ramis, with Bill Murray, Andie MacDowell, and Chris Elliott (2001).

- *Crash,* directed by Paul Haggis, with Sandra Bullock, Don Cheadle, Matt Dillon, Terrence Howard, and Chris Bridges (2005).
- *Eternal Sunshine of the Spotless Mind,* directed by Michel Gondry, with Jim Carrey, Kate Winslet, and Tom Wilkinson (2004).
- *Rain Man,* directed by Barry Levinson, with Dustin Hoffman, Tom Cruise, and Valeria Golino (1988).

Chapter 4: Recognizing the People We Know

Our ability to notice other people is sometimes shockingly poor. We normally recognize family and friends quickly, but we can have trouble recognizing briefly seen strangers, as evidenced by cases of mistaken identity. Yet, there are occasions when recognizing a familiar person is not easy. What would it take for an imposter to fool us? For those we know well, are they defined solely by their physical characteristics, or might psychological factors including their personality and memory also play key roles in identification? Problems can arise when we take others at face value.

Films Involving Person Recognition

- *The Imposter,* docudrama directed by Bart Layton, with Frédéric Bourdin (2012).
- *The Return of Martin Guerre,* directed by Daniel Vigne, with Gérard Depardieu and Nathalie Baye (1982, French with English subtitles).
- *Face/Off,* directed by John Woo, with Nicholas Cage and John Travolta (1997).
- *Invasion of the Body Snatchers,* directed by Philip Kaufman, with Donald Sutherland and Brooke Adams (1978).
- *Changeling,* directed by Clint Eastwood, with Angelina Jolie and John Malkovich (2008).

Chapter 5: Autobiographical Memories and Life Stories

Autobiographical memory entails a myriad of personal experiences that combine to form a life story. Whether these memory narratives involve events such as a first day at school, a first crush, or the loss of a loved one, they are often colored by emotion, and experiences that touch us emotionally tend to be remembered well. In telling our life stories, we function like authors of an autobiography, taking snippets of different experiences and weaving them into a coherent narrative.

Films Involving Autobiographical Memory

• *The Kite Runner*, directed by Marc Foster, with Khalid Abdalla, and Zekira Ebrahimi (2007, English and Dari with English subtitles).

• *Slumdog Millionaire*, directed by Danny Boyle, with Dev Patel, Freida Pinto, and Madhur Mittah (2007).

• *Cinema Paradiso: The New Version*, directed by Giuseppe Tornatore, with Salvatore Cascio, Philippe Noiret, and Jacques Perrin (2002, Italian with English subtitles).

• *Titanic*, directed by James Cameron, with Kate Winslet and Leonardo DiCaprio (1997).

• *Unchained Memories: Readings from the Slave Narratives*, documentary directed by Ed Bell and Thomas Lennon (2003).

• *The Joy Luck Club*, directed by Wayne Wang, with Ming-Na Wen, Tsai Chin, Rosalind Chao, and Tamlyn Tomita (1993, English and Mandarin with English subtitles).

Chapter 6: When Troubling Memories Persist

Many everyday memories fade quickly, whereas traumatic memories can linger for years. Combat veterans can suffer from painful memories, leading to *post-traumatic stress disorder*, a debilitating condition in which they continually relive their battlefield experiences and find it hard to cope with everyday stresses. Although any life-threatening event can produce a long-lasting traumatic memory, many of these disturbing memories can be effectively treated.

Films Involving Troubling Memories

• *Remembrance,* directed by Anna Justice, with Dagmar Manzel, Alice Dwyer, and Mateusz Damięcki (2011, English, German, and Polish with English subtitles).

• *Born on the Fourth of July,* directed by Oliver Stone, with Tom Cruise, Frank Whaley, Willem Dafoe, and Kyra Sedgwick (1989).

• *Rachel Getting Married,* directed by Jonathan Demme, with Anne Hathaway, Rosemarie DeWitt, Bill Irwin, and Debra Winger (2008).

• *Mystic River,* directed by Clint Eastwood, with Sean Penn, Tim Robbins, and Kevin Bacon (2003).

• *Capturing the Friedmans,* documentary directed by Andrew Jarecki, with Arnold Friedman and family (2003).

• *Ordinary People,* directed by Robert Redford, with Timothy Hutton, Donald Sutherland, Mary Tyler Moore, and Judd Hirsch (1980).

Chapter 7: Understanding the Reality of Amnesia

Amnesia is the most common memory topic in film, and it is the one that filmmakers most often get wrong. Characters suffer memory loss after a blow to the head in one scene, only to recover their memory in a later scene, following a second head bonk. In fact, concussions can lead to amnesia, but they never produce memory recovery. Memory problems can occur following neurological damage or upsetting psychological experiences, with some types of amnesia being permanent and others showing recovery.

Films Involving Amnesia

- *Desperately Seeking Susan,* directed by Susan Seidelman, with Rosanna Arquette, Aidan Quinn, and Madonna (1985).
- *The Music Never Stopped,* directed by Jim Kohlberg, with Lou Taylor Pucci, J. K. Simmons, and Julia Ormond (2011).
- *The Return of the Soldier,* directed by Alan Bridges, with Julie Christie, Glenda Jackson, Ann-Margaret, and Alan Bates, (1982).
- *The Majestic,* directed by Frank Darabont, with Jim Carrey and Martin Landau (2001).
- *Primal Fear,* directed by Gregory Hoblit, with Richard Gere, Edward Norton, and Laura Linney (1996).
- *Memento,* directed by Christopher Nolan, with Guy Pearce, Joe Pantoliano, and Carrie-Anne Moss (2000).

Chapter 8: Senior Moments, Forgetfulness, and Dementia

Elderly people know that their memory changes with age and that few of these changes are improvements. There are normal changes in memory that all seniors typically experience, and there are debilitating pathological changes that increasingly victimize people in their later years. Some changes, as in a declining ability to recall names, are normal, whereas others, such as forgetting the way home, could signal a more serious decline due to the onset of Alzheimer's disease.

Films Involving Memory in Old Age

- *Cocoon,* directed by Ron Howard, with Don Ameche, Wilford Brimley, Hume Cronyn, Maureen Stapleton, Jessica Tandy, and Gwen Verdon (1985).

• *On Golden Pond,* directed by Mark Rydell, with Henry Fonda, Katherine Hepburn, and Jane Fonda (1981).

• *Amour,* directed by Michael Haneke, with Jean-Louis Trintignant and Emmanuelle Riva (2013, French with English subtitles).

• *Away from Her,* directed by Sarah Polley, with Julie Christie, Gordon Pinsent, and Olympia Dukakis (2006).

• *The Bucket List,* directed by Rob Reiner, with Jack Nicholson and Morgan Freeman (2007).

Fade-Out

All of the films in this book say something about our experience of memory and the role that it plays in our lives. Watching these films will add to the richness of this experience. For some chapters, one or two films may be sufficient for viewing, whereas for other chapters, comparing more films might be more informative. To get the most from this book, read a chapter before making your film selections. Watch the films for enjoyment, but watch them with an educated eye.

2 Movies and the Mind's Workbench

Setting the Scene

Working memory has been called the mind's workbench. Filmmakers employ it to create movies and viewers use it to experience them. This temporary memory system enables us to maintain a relatively small number of thoughts for a short period of time. As soon as our attention is drawn elsewhere, those thoughts are quickly forgotten. Thinking about the present, remembering the past, and imagining the future depend on working memory processes. Although no movies depict working memory per se, without it, we could not have film.

Featured Films

Fantasia, animation from Walt Disney Studios (RKO Radio Pictures, 1940, Rated G).

Life of Pi, directed by Ang Lee, with Suraj Sharma, Irrfan Khan, and Ayush Tandon (20th Century Fox, 2012, Rated PG).

Pulp Fiction, directed by Quentin Tarantino, with Samuel L. Jackson, John Travolta, and Uma Thurman (Miramax Films, 1994, Rated R).

Mulholland Drive, directed by David Lynch, with Naomi Watts, Laura Harring, and Justin Theroux (Universal Pictures, 2001, Rated R).

Avatar, directed by James Cameron, with Sam Worthington, Zoë Saldana, and Sigourney Weaver (20th Century Fox, 2009, Rated PG-13).

Engineering the Imagination

Can you imagine a purple cow? You can by engaging your mind's workbench to think about cows and colors. Purple cows, says writer Priscilla Long, require memory:

> How could I imagine a purple cow, if I could not remember the cows of my childhood switching their tails against horseflies? How could I imagine a purple cow if I could not remember purple crayons, purple potatoes, purple grape juice?[1]

Generated by the mind's workbench, creative images enable filmmakers to tell us fanciful tales—the type of imaginative thinking made visible long ago in Walt Disney's film *Fantasia*.

Fantasia was a 1940 experiment in animation in which Disney took various symphonic pieces and created artwork to illustrate the musical scores. Disney called his animators *imagineers*—engineers of the imagination—and they were not bound by the composers' original intent when they animated their musical selections. Sometimes their animation is abstract, as in Beethoven's Symphony no. 5 in C Minor, whereas other times it is highly concrete, as in their use of Mickey Mouse in Dukas's *The Sorcerer's Apprentice*. But regardless of visual style, the animated musical selections are candy for the eye and ear. Today, filmmakers use computer-generated imagery to create characters that are so realistic that we have difficulty distinguishing the real from the artificial on the screen.

"So which story do you prefer?"
Life of Pi

Based on Yann Martell's book, Ang Lee's film *Life of Pi*, featuring Suraj Sharma, Irrfan Khan, and Ayush Tandon, tells the story of a sixteen-year-old Indian boy named Piscine Patel, who goes by the name of Pi to avoid teasing by his classmates. After losing his family in a shipwreck, Pi drifts across the Pacific in a lifeboat, accompanied by a Bengal tiger named Richard Parker. Lee's 3D computer-generated images transform Martell's story into a stunning sensory experience where we can almost feel the softness of the tiger's fur and taste the raindrops falling from a passing shower; we see one dazzling image after another, including a phosphorescent whale leaping high into a starlit sky.

Exhausted after 227 days at sea, Pi washes ashore on a sandy Mexican beach, only to watch his seafaring companion wander off without so much as a glance back in his direction. Later, recovering in a hospital bed, Pi describes his adventure to investigators. When they express skepticism over his story, Pi offers them a second, more acceptable version. Flashing forward in time, the adult Pi finishes telling his tale to a writer.

> Pi: I've told you two stories about what happened on the ocean. Neither explains what caused the sinking of the ship, and no one can prove which story is true and which is not. ... So which story do you prefer?
> Writer: The story with the tiger. That's the better story.

In the end, we are offered two interpretations of Pi's story, and it is up to us to decide which version was real, even though, as viewers, we never question the reality of the computer-generated characters and scenes.

In watching *Fantasia* or *Life of Pi*, we get to see what people with extraordinary imaginations can do when they are let loose on their drawing boards or computers. Their highly creative images relied largely on their ability to interpret and imagine, processes that draw heavily on working memory.

The Long and the Short of Memory

Short-term and long-term memories differ in capacity and duration. *Short-term memory* refers to our ability to retain a small amount of information for a short period of time; *long-term memory* involves our ability to retain a vast amount of information for an indefinite period of time. Processes involving episodic, semantic, and procedural memory determine long-term retention, whereas processes used to maintain and manipulate thoughts in conscious awareness influence short-term retention. Sometimes filmmakers get these processes confused, as when characters in Disney's *Finding Nemo* and Christopher Nolan's *Memento* state "I have no short-term memory," in referring to their inability to recall the recent past. For psychologists, remembering what you did yesterday or a few hours ago is recent long-term, not short-term, retention. Short-term memory is what is on your mind at this moment, and the memory system responsible for short-term retention is called *working memory*. Essential for all of our mental activity, we use working memory to think verbally and nonverbally—whether texting a friend, drawing a picture, or daydreaming about some future event—and it is hard to overstate its importance.

Yet, because working memory is temporary, it is particularly prone to forgetting and frustration. For instance, before the advent of the Global Positioning System (GPS), I would sometimes ask a stranger for driving directions that, too often, went something like this:

"Go down this road, turn left at the first light, continue until you come to a stop sign, turn right and continue for three blocks, then take a sharp left across the railroad tracks and look for a large brick building on the right. You can't miss it."

After which I thanked the person, drove off, and then thought to myself:

"OK, I continue down this road and take a left at the first light ... (pause) ... or was it a right at the first light? Hmm. I am still lost."

I heard the directions and understood them, but I could not keep them in mind because they exceeded my working memory capacity. Working memory enables us to keep a few items in mind, typically for a matter of seconds. When there is too much to think about or it must be maintained too long, the information is quickly forgotten. This transient system can hold approximately seven items for several seconds, unless we keep refreshing those items by mentally rehearsing them. As soon as our attention is drawn elsewhere, we quickly forget. In my driving example, there were just too many landmarks and street turns to maintain.

But forgetting in working memory is not always the result of trying to think about too much. Sometimes we forget when we are only thinking of one thing. All of us have had the frustrating experience of walking from the living room to the bedroom to retrieve some item, only to forget what we were looking for once we arrived. This type of absent-minded forgetting occurs when we let our thoughts wander while walking, because new thoughts can knock old thoughts out of conscious awareness.

New thoughts can also prevent other thoughts from entering conscious awareness, as when we think about an upcoming event while tossing our car keys down, only to wonder later, "Now, what did I do with those keys?" This type of annoying forgetting is pervasive but paradoxically valuable, because we would likely be overwhelmed with useless clutter if we permanently remembered every action or thought, such as every time we went to the bedroom to get something.[2] Sometimes we need our memory to last just long enough to complete a desired task. Working memory serves this temporary function admirably, and it does so in a variety of ways.

Thought's Three Components

Maintaining and manipulating thoughts in working memory, says psychologist Alan Baddeley, involves the joint operation of three components—the *central executive, phonological loop*, and *sketchpad*—each serving a different mental function.[3] With links to our episodic memory, these components normally perform in a seamless, integrated manner to influence our conscious awareness.

The central executive acts as a corporate supervisor, planning and coordinating the activity of the other components. It allocates our limited attention—*externally* to stimuli striking our senses or *internally* to thoughts we may be mulling over—through neural processes involving the frontal lobes of the brain.

The phonological loop is the verbal component of working memory. It can maintain a limited number of speech sounds for a short period of time through processes involving our left frontal and left temporal lobes. When we need to remember a phone number, for instance, we engage the phonological loop to maintain the number temporarily by mentally rehearsing or "looping" it until we can enter it into our cell phone.

Finally, drawing on neural processes involving our temporal, parietal, and occipital lobes, the sketchpad is the visual component of working memory. The sketchpad enables us to generate and manipulate visual images—as when we imagine purple cows or mentally rearrange our living room furniture. Brain areas that are important for visual perception are also important for visual imagery, as seeing and imagining share some of the same neural networks.[4]

Taking a Mental Walk

The sketchpad does not work in isolation. It relies on our long-term memory and the other working memory components to generate its visual images. To see how, try the following task:

Close your eyes and mentally count the number of windows in your home. Once you have a total, continue reading.

Keeping your window count in mind, consider the mental operations that you used to reach that number. First, you had to access your long-term episodic memory to retrieve a plan of your home. This plan enabled you to generate a series of visual images depicting each room using your working memory sketchpad. Visualizing each of those images allowed you to count the windows and continually update that number in your phonological loop ("1, 2, 3," and so forth). Finally, your central executive directed the operation of these components as you mentally walked through your home, counting windows. By accessing information already in long-term memory, your working memory produced the conscious experience of counting imaginary windows, seen with your mind's eye.

Memory's Role in Watching Films

We need memory while watching films to interpret and understand what we see. We engage our memory systems to understand individual scenes and connect those scenes into a coherent story. We need semantic memory to understand a film's dialogue and action, episodic memory to remember the sequence of events being shown, and working memory to focus on individual scenes and make appropriate connections between them. Without memory, we would not have stories; without stories, we would not have film. Good stories have a transporting quality; they grab our attention and mentally transport us into the narrative world of the story.[5]

Whether watching a film or reading a book, we engage our memory systems to understand those stories that capture our attention—stories about people trying to achieve a difficult goal and how they are changed by the experience. A story, says writer Lisa Cron, is not just about something that happens. Stories grab us, she says, when they show us how people are changed by a difficult experience, thereby allowing us to imagine how we would feel if presented with a similar challenge.[6] This ability to empathize with a story's character, to imagine ourselves in a similar situation, is memory's essential contribution to understanding the films that we watch.

Film watching also makes demands on our conscious attention; demands that are heavily influenced by a film's plot structure. Simply structured films that employ a single plot unfolding chronologically over time are easy to follow because they make little continuous demand on attention. Maintaining a story line is easy when films follow this linear structure: We meet the main characters and observe the protagonist dealing with an incident in the first act; the protagonist's attempts to deal with the incident often produce worsening conditions in the second act; and, finally, the protagonist's struggles resolve the incident leading to a climax in the third act, often followed by an epilogue.[7]

Breaking with this format, complexly structured films can involve multiple plots with time lines that double back on themselves, demanding considerable attention to follow the film. *Pulp Fiction*, featuring Samuel L. Jackson, John Travolta, and Uma Thurman, is one notable example. This film employs a nonlinear, unconventional structure that fully demands our attention, but it is never confusing because writer and director Quentin Tarantino help us understand it by reminders that repeatedly engage our memory.

"You know what they call a Quarter Pounder
with Cheese in Paris?"
Pulp Fiction

This complexly structured film is an anthology of crime stories involving a collection of surprisingly interesting lowlifes. Beginning and ending in the same diner, *Pulp Fiction* tells three different but related stories, shows scenes out of sequence, and jumps back and forth between stories—yet we are never lost. Our memory systems hold this film together to reveal an underlying continuity through the filmmaker's engaging dialogue and action.

Pulp Fiction packages three familiar crime story themes into a trilogy to show what can happen when events cause the story lines to unravel.[8] The first story involves a gangster who must serve as an escort for his boss's attractive wife. The second story involves a boxer who is ordered to throw a fight, but refuses. The last story shows what happens when two professional hit men accidently shoot someone in their car. Because the same characters move in and out of each story, we experience the three stories as one. What is fascinating about *Pulp Fiction* is how well its seemingly irrelevant dialogue sticks in the viewer's memory, bridging the different scenes in the film.

Early on, we observe Jules and Vincent, two hit men en route to a job to retrieve a mysterious briefcase stolen from their boss. While driving, the men engage in what appears to be incongruous banter, but, in reality, their dialogue is preparing viewers for various scenes that will subsequently appear in the film. At one point they talk about hamburgers and what people in France call a Quarter Pounder with Cheese.

> Vincent: You know what they call a Quarter Pounder with Cheese in Paris?
> Jules: They don't call it a Quarter Pounder with Cheese?
> Vincent: No man. They got the metric system there ...
> Jules: What do they call it?
> Vincent: They call it a Royale with Cheese.

Arriving at the apartment of the young men who stole the briefcase, the killers find Brett, the leader of the gang, eating a hamburger with his friends. The dialogue grabs our attention because we remember the earlier hamburger discussion.

> Jules: Whatcha having?
> Brett: Hamburgers.
> Jules: You know what they call a Quarter Pounder with Cheese in France?
> Brett: No.
> Jules: Tell him Vincent.
> Vincent: Royale with Cheese.
> Jules: Know why they call it that?
> Brett: Umm, because of the metric system?
> Jules: Check out the big brain on Brett!

Pulp Fiction should be difficult to follow, but it is not because this film actively engages our episodic and working memory by bringing up ideas in one scene, only to return to them in later scenes—such as when a character starts a joke about tomatoes in one scene and gives the punch line in another. The use of multiple visual and verbal reminders serves to keep the story elements fresh in a viewer's memory and connect the various characters in a crime drama that appears at once both familiar and entirely unique. *Pulp Fiction* is absorbing and sometimes violent, but it is never boring, and, oddly, it never seems repetitious. Through its effective use of memory, *Pulp Fiction* makes a complex story understandable by keeping our attention glued to the film.

Attention, Multitasking, and Confusion

"Everyone knows what attention is," wrote psychologist and philosopher William James in 1890.[9] "It is the taking possession by the mind, in clear and vivid form, of one out of what seem several simultaneous possible objects or trains of thought." In reading these lines, numerous stimuli compete for your attention. There may be visual stimuli from people nearby, auditory stimuli from their conversations, and tactile stimuli on different parts of your body. Rather than being overwhelmed with these sensory inputs, the central executive comes to your aid by focusing on the written words, while tuning out all of the rest. This need to focus occurs because our limited supply of attention is often subject to competing demands.

Normally, when we are given two attention-demanding tasks to do at the same time—such as playing a piano while repeating spoken words—our performance on one or both tasks suffers compared to our performance on either task alone. However, after extensive practice, people can successfully multitask by performing these tasks concurrently. They can read a story or play a piano while repeating dictated words almost as accurately as they can perform each task alone.[10] Most of us have difficulty with multitasking when activities divide our attention, such as using a cell phone while driving. Yet, even here, a few notable individuals have demonstrated supernormal multitasking ability.

These people, called *supertaskers*, can verify math equations and remember a series of words spoken on a cell phone (e.g., "Is [3/1]–1 = 2?—Cat—Is [2 × 2] + 1 = 4?—Box") before recalling the words while behind the wheel in a driving simulator on a busy highway. Supertaskers perform the driving and math tasks together as well as they do each one alone.[11] This special ability has been found in only a few individuals to date. If you wonder whether you too might be a supertasker, the odds are stacked heavily against you—exceptional mental performance is always rare. More often, when events exceed our attentional limits, we experience confusion and cognitive disruption, feelings that viewers frequently report while watching David Lynch's *Mulholland Drive*. This complexly structured, attention-demanding film, featuring Naomi Watts, Laura Harring, and Justin Theroux, resists our attempts at finding coherence, existing instead as a sequence of transient, dreamlike experiences.

"Hey, pretty girl. Time to wake up!"
Mulholland Drive

Mulholland Drive is perplexing because it seems to disengage all of our memory systems save one and force us to experience it solely with working memory. We experience a sequence of scenes that each makes sense on its own, but there is no obvious story line to connect the individual scenes as our attention shifts from one to the next. Characters appear and reappear at different points, and sometimes we are left wondering if the characters themselves are the same. Watching this film is like eavesdropping on someone's dreams—each one understandable alone, but each one quickly forgotten as soon as the next dream comes along. Even the opening scene of someone falling face down into a pillow hints that a dream is about to begin.

Following the opening scene, *Mulholland Drive* shifts in film noir fashion to the road of the same name, at night in the Hollywood hills. Alone in the back seat of a car and facing a man with a gun, a beautiful woman is about to be shot, when, suddenly, she is thrown from the car by a crash and develops amnesia. Stumbling into the apartment of an aspiring actress named Betty, the woman with no memory spots a movie poster of Rita Hayworth and tells Betty that her name is Rita. Together, they set out to determine Rita's identity, starting with her purse containing $250,000 in cash and a strange blue key.

The amnesic Rita seems wise to the ways of Hollywood, whereas the newly arrived Betty pines for fortune and fame in the land of dreams, saying, "Of course, I'd rather be a great actress than a movie star. But, you know, sometimes people end up being both … I am just so excited to be here. I mean, I just came here from Deep River, Ontario, and now I am in this dream place." This line may be a clue that Betty is literally dreaming and that viewers of *Mulholland Drive* are watching her dreams as she sleeps. Each dream is like a sketchpad image that is held temporarily in working memory until the next one bumps it out. Our working memory is fully engaged in watching *Mulholland Drive*—focusing on one scene after another—just as the contents of our conscious awareness can change from moment to moment. But without help from our other memory systems, we cannot connect the sequence of scenes and form an understanding of this film. Much like our inability to connect the sequence of dreams that we experience each night, *Mulholland Drive* shows the mental confusion that would result if working memory were our only memory system.

The idea that *Mulholland Drive* may be nothing more than the stuff of dreams is one interpretation. David Lynch has not revealed his film's meaning, saying that viewers must form their own interpretation. Unlike *Pulp Fiction* where verbal and visual reminders help viewers connect different story lines, *Mulholland Drive* offers no viewing assistance. For many viewers, this film presents a puzzle that is too complex for their working memory to solve all alone.

Working Memory Problems

Pulp Fiction and *Mulholland Drive* illustrate how filmmakers can use working memory to influence our film-viewing experience in fundamentally different ways, just as it can influence our conscious experience in everyday life. When one or another of the working memory components breaks down, our real-life experiences can be radically altered.

Planning with the Central Executive

Thinking about the present and imagining the future are important central executive functions involving the brain's frontal lobes. When there is brain damage in this area, our ability to plan and imagine is compromised, as shown by the strange tale of Phineas Gage.

In 1848, a construction crew, led by their likeable foreman—twenty-five-year-old Phineas Gage—was blasting rock in rural Vermont. The workers had poured blasting powder into a hole and Gage was bent over the hole, tamping the gunpowder down with a heavy iron bar. While tamping the blasting powder, the bar struck a stone and sparked. Unfortunately for Gage, his head was right over the hole when the gunpowder exploded. The force of the explosion sent the iron bar cleanly through Gage's head. Incredibly, he survived this terrible cerebral assault, and recent computer-aided reconstructive scans of his skull indicate that his left frontal lobe was extensively damaged.[12] After a few weeks recuperation, Gage was again up and about, but there was now a profound change in his behavior. Said the physician who attended him:

> He is fitful, irreverent ... impatient of restraint or advice when it conflicts with his desires, yet capricious and vacillating, devising many plans for future operation, which are no sooner arranged than they are abandoned... his friends and acquaintances said that he "was no longer Gage."[13]

This case might be just a neurological oddity were it not for the fact that more recent patients with frontal lobe damage show similar problems in planning and social behavior.[14] Our frontal lobes are important, says neuroscientist Antonio Damasio, for anticipating future events and making appropriate plans.[15] Working memory's central executive—the component that was damaged in Gage—guides this planning process.

Rehearsing with the Phonological Loop

Besides formulating plans, the central executive directs the phonological loop to maintain speech sounds in conscious awareness. When a friend speaks, we hear a sequence of sounds that we segment into words, and we hold these words in our phonological loop long enough to grasp their meaning from connections to our semantic memory. With each new utterance, the previously heard words are discarded, and new words are processed for meaning. This temporary storage process is continually repeated, enabling us to grasp the meaning of our friend's speech.

Normally, we can only hold a limited number of speech sounds temporarily in the phonological loop, but sometimes we can increase it to surprising lengths. To see how, read the following list of nine words, and then try to recite them in order.

Lincoln, Milky, Criminal, Differential, Address, Way, Lawyer, Calculus, Gettysburg

Most people have a hard time recalling all of the words because they exceed our memory span of approximately seven items. The span is the number of items we can accurately maintain in our phonological loop, and this capacity is severely reduced if a person suffers damage to the speech areas of the left hemisphere. One patient, KF, suffered this type of damage from a motorcycle accident and, while able to speak, he was unable to recall more than two digits. If you spoke the three-digit sequence "7, 3, 8" to him, he would be unable to repeat it.[16]

Normal working memory capacity is not so severely restricted, and it can be extended. Read the following four phrases and try recalling them in order.

Lincoln's Gettysburg Address, Milky Way, Criminal Lawyer, Differential Calculus

This time you likely recalled more of these words than on your previous attempt. The words are the same as before, but now they are organized into four phrases. Psychologist George Miller referred to these organized units as *chunks* of information.[17] Given enough practice, we can even chunk long strings of digits. One person, SF, was tested on digit strings for many days. During the first week, he had a typical span of seven digits, but after a year's practice, he could recall eighty digits in order following a single exposure. SF made a game out of this task by turning the numbers into meaningful units. For example, he thought of the sequence "8, 9, 3" as "89 point 3, a very old man."[18]

Although we can use chunking strategies, most of the time we will retain roughly seven items in working memory, a limit that is largely determined by the speed with which we can pronounce the speech units. For instance, which list of five words do you think will be easier to recall?

Boy, Cow, Desk, Shoe, Truck

Archaeology, Electricity, Hippopotamus, Refrigerator, University

The list of single-syllable words is the clear winner because we can rehearse its speech sounds faster than those from the multisyllable list. Typically, our phonological loop is limited to the number of items that we can pronounce in 1.5 seconds.[19] How long we retain these items depends on how long we keep rehearsing them. Once rehearsal stops, those speech sounds are forgotten in a matter of seconds.[20]

The phonological loop is essential for language acquisition, a process that begins during early childhood when children rapidly build up their vocabulary. This component enables children to hold on to the sounds of new words so that they can repeat them and learn their meaning. Children who do poorly on memory span tasks tend to have difficulty mastering vocabulary, whereas youngsters who do well on these tasks excel at vocabulary acquisition.[21]

Imagining with the Sketchpad

Complementing the phonological loop is the sketchpad. We can think in words with our phonological loop and in mental pictures with our sketchpad. For example, try naming the color of the stars on the American flag. To get the answer, you most likely generated a mental image of the flag and "saw" that the stars are white. You would likely do the same thing if I asked you whether a kangaroo has a short or a long tail by imagining a kangaroo and noting that its tail is long. These tasks are easy because we can retrieve information about the appearance of flags and kangaroos from our long-term memory and examine those images in our sketchpad.

Our ability to form mental images depends on two visual systems in the brain, technically called the *parvocellular* and *magnocellular* systems.[22] These systems normally function well together, but damage to either system could impair our ability to imagine what an object looks like or describe where an object is located. People with brain damage to one or the other of these systems reveal strangely altered visual experiences because their sketchpads no longer function normally.

Brain damage in the occipital and temporal lobes that contain the parvocellular system can lead to a disorder called *object agnosia*, a difficulty in identifying common objects such as a comb or spoon. One patient, LH, suffered brain damage in these lobes from a car crash and had trouble identifying common objects from memory. He could copy a drawing of a pencil or teabag, for example, but had difficulty naming his drawings.[23]

Brain damage to the occipital and parietal lobes that house the magnocellular system can lead to a different neurological disorder called *visual neglect*. People with this type of damage, especially in their right parietal lobe, largely ignore the left side of their visual world. They may eat only from the right side of a dish, comb only the right side of their hair, and copy only the right side of pictures. Yet, these patients do not have a problem with their eyes. If you point out the missing left half of a drawing, for instance, neglect patients will see that it is missing, saying, "Oh, I didn't notice. I don't know how I could have missed it." Then they go right back to ignoring their left side. Fortunately, rehabilitation training can help these patients learn to refocus their attention on both sides.[24]

These dysfunctions illustrate how damage to the sketchpad's neural networks can alter our visual experience. With a normally functioning sketchpad, we have no trouble imagining the color of a banana or the left side of the street that we live on. We can even be visually creative and imagine things never seen. Our sketchpad affords us this mental flexibility, flexibility that filmmakers have used in creating imaginary worlds. James Cameron's *Avatar*, featuring Sam Worthington, Zoë Saldana, and Sigourney Weaver, is a marvel of this type of imaginative thinking. It transports us into the future, to a distant planet where a young marine embarks on the mythic journey of the hero.

"One life ends, another begins."
Avatar

Mythic stories have been passed down for ages, from every continent on Earth. Common to many cultures, says mythologist Joseph Campbell, is the myth of the hero's journey. Regardless of where this story is told, it reveals the same underlying structure:

> A hero ventures forth from the world of common day into a region of supernatural wonder: fabulous forces are there encountered and a decisive victory is won: the hero comes back from this mysterious adventure with the power to bestow boons on his fellow man.[25]

Many films have depicted a reluctant hero taking up an arduous journey with characters as varied as Luke Skywalker in George Lucas's *Star Wars*, John Dunbar in Kevin Costner's *Dances with Wolves*, and even Simba in Disney's *The Lion King*. But for Jake Sully's journey in *Avatar*, writer and director James Cameron constructed an entire new world, populated by a race of hunters and gatherers who, reminiscent of Native Americans, strive to live in harmony with nature.

In 2154, with the Earth's natural resources rapidly becoming depleted, a mining company, seeking a valuable mineral, sends an expeditionary military force to mine the densely forested planet of Pandora. Standing in their way are the Na'vi people whose gathering place is a giant hardwood called Hometree, under which lies the rich mineral deposit. The Na'vi are ten feet tall, blue skinned and golden eyed, with a prehensile tail.

Jake Sully is a paraplegic marine who, after promised surgically restored legs, attempts to persuade the Na'vi to move. He does so by using an avatar, a Na'vi lookalike that Jake controls, wired up from his home base. As an avatar, Jake not only walks again, he fully experiences Pandora as a Na'vi, venturing forth into its Eden-like forest. But Jake soon learns that the forest is dangerous, and he finds himself fending off hungry predators, until a female Na'vi rescues him. Her name is Neytiri, and she is contemptuous of Jake's ignorance, after slaying an animal to save him.

> Neytiri: You have a strong heart. No fear. But stupid! Ignorant like a child!
> Jake: Well, if I'm like a child, then look, maybe you should teach me.
> Neytiri: Sky People cannot learn, you do not see.
> Jake: Then teach me to see.
> Neytiri: No one can teach you to see.

For the Na'vi, seeing is akin to understanding, and Jake has much to learn about the Na'vi's deep connection to the forest and its creatures. Reluctantly, Neytiri becomes his mentor and more. As they fall in love, Jake gradually adopts the Na'vi way of life, becomes a warrior and leader, and fights off the destructive mercenaries as they advance toward the mineral deposits. After an epic battle, with the Na'vi and their related clans victorious, Neytiri saves Jake's life one final time, as his human form struggles for oxygen. Recovering, he says, "I see you." His former life ended, his new one begins. With Jake permanently transformed into his avatar, Campbell's mythic sequence is complete. Jake heard the call to adventure, received supernatural aid from a mentor, met a goddess, and triumphantly saved the Na'vi people. Jake completed the hero's journey, but only with Neytiri by his side. She is every bit the hero as Jake.

Avatar combines mythic storytelling and attention to detail with unbelievably engaging visual imagery. There are so many stunning images, but those of Neytiri and Jake flying through Pandora's floating mountains are breathtaking. In watching *Avatar*, we are seeing approximately 60 percent computer-generated images and 40 percent live action, but the imagery is so realistic, right down to the facial movements of the avatars, that we become fully captured by the fantasy.[26] It is our working memory that enables us to "see" it.

Fade-Out

Working memory enables us to maintain and manipulate concepts and images in conscious awareness. It is essential for thinking about the past, concentrating on the present, and imagining the future—qualities that were demonstrated in various ways and to varying degrees by the films in this chapter. Watching a film mimics our use of memory in everyday life, for without the use of our different memory systems, and most especially our working memory, film viewing would be an empty experience.

The films in this chapter illustrate how filmmakers and viewers rely on working memory to create and experience film. The films in remaining chapters will provide rich examples of how diverse memory phenomena influence our everyday behavior.

3 Making Memories That Last

Setting the Scene

Effective learning and remembering enables us to imagine and plan the future—the primary function of memory. Creating long-lasting memories involves attending to the meaning of an item or event and relating it to information already in memory. These meaningful associations enhance our ability to remember. Playing on this theme, the films in this chapter show how the past influences the present and how remembering is necessary for future thinking.

Featured Films

After Life, directed by Hirokazu Kore-eda, with Arata, Erika Oda, and Taketoshi Naito (Artistic License, 1998, Japanese with English subtitles, Unrated).

Cast Away, directed by Robert Zemeckis, with Tom Hanks and Helen Hunt (20th Century Fox, 2000, Rated PG-13).

Groundhog Day, directed by Harold Ramis, with Bill Murray, Andie MacDowell, and Chris Elliott (Columbia Pictures, 2001, Rated PG).

Crash, directed by Paul Haggis, with Sandra Bullock, Don Cheadle, Matt Dillon, Terrence Howard, and Chris Bridges (Lionsgate, 2005, Rated R).

Eternal Sunshine of the Spotless Mind, directed by Michel Gondry, with Jim Carrey, Kate Winslet, and Tom Wilkinson (Focus Features, 2004, Rated R).

Rain Man, directed by Barry Levinson, with Dustin Hoffman, Tom Cruise, and Valeria Golino (United Artists, 1988, Rated R).

A Memory Worth Keeping

The memories that we hold dear tell us something about ourselves. Think for a moment about your happiest times. Recalling those events can bring forth a quick smile and sense of elation.[1] We have a sense of recapturing the past, feeling what we felt before, yet knowing that we are remembering. Encoding the present and retrieving the past are based on *episodic memory,* our memory of past experiences. To understand the power of these experiences, writer and director Hirokazu Kore-eda asked five hundred people for their most cherished memory. From these interviews, he created *After Life,* a documentary-style, celestial fantasy to highlight the life-defining quality of these memories.

> "As soon as you've relived your memory you will move on, taking only that memory with you."
>
> *After Life*

Kore-eda's film, featuring Arata, Erika Oda, and Taketoshi Naito, introduces us to twenty-two people, some young and some old, who recently died. One by one, they arrive at a drab, cosmic way station, somewhere between Heaven and Earth, where they must complete a critical task before continuing on to the afterlife. Caseworkers assigned to the deceased relay their assignment:

> You'll be staying with us for one week ... While you are here, there is one thing you must do ... We need you to select one memory. One memory that was most meaningful or precious to you.

The deceased have three days to select the single memory that they will keep for eternity; all other memories will be lost.

Once a memory is selected, the caseworkers film a recreation of the event and show it to the deceased—mimicking how our memories are interpretations of past experiences, not exact replicas of those events. Departing with that recreated memory, the deceased will reexperience it forever. Surprisingly, the selected memories are often brief snippets of time, evoking subtle pleasurable sensations. A former pilot remembers flying through wispy clouds, while an elderly woman recalls cherry blossoms falling around her. For some, however, recalling happiness is no easy task.

Failing to find a happy memory while looking back on his dull life with regret, elderly Ichiro Watanabe finally settles for an uneventful memory—one of him sitting quietly with his wife, Kyoko. But for his youthful caseworker, Mochizuki, this is an eye-opening moment. Kyoko was the woman Mochizuki was engaged to marry before he was killed in the Pacific War. Unable to find a happy memory of his own, Mochizuki became a caseworker, stuck in the way station for years.

It is only after a fellow caseworker shows him the favorite memory of Ichiro's deceased wife Kyoko, saying, "She chose this moment," does he experience joy. Looking on, Mochizuki sees that Kyoko's happiest memory shows her sitting on a park bench with him, a young naval officer in 1943. Acknowledging this discovery as his happiest moment, Mochizuki is prepared to move on, adding, "I searched desperately inside myself for any memory of happiness. Now, fifty years later, I've learned I was part of someone else's happiness. What a wonderful discovery."

For filmmaker Kore-eda, memories give our life meaning, and heaven is our happiest memory. Given the same task, which memory would you treasure most and wish to maintain forever?

Creating Long-Lasting Memories

Fortunately, we are not limited to a single memory; we can remember many experiences over our lifetime, and return to them again and again. The key to forming these long-lasting memories is understanding the role of attention.

Common Cents about Memory

To appreciate the importance of attention, try drawing a penny or other common coin from memory—an object that you have probably seen hundreds of times. This task should be easy, not hard. Yet, when people were asked to draw a penny from memory, numerous errors occurred. People knew the color and size of the coin, and they remembered that Lincoln's head was shown, but they had him looking left or right equally often, they frequently left off the word *Liberty*, and they put features such as the date and *E Pluribus Unum* in the wrong locations. This problem was not caused by an inability to draw. When people were shown an array of twelve pennies consisting of one correct penny and eleven pennies with an incorrect feature, they still had a tough time recognizing the actual coin.[2] Recognition is poor because we normally give minimal attention to the many features engraved on a penny. Faced with a handful of coins, a quick glance at their color and size is sufficient to select a penny from the nickels and dimes. In this instance, limited attention to detail leads to limited memory.

What type of attention is needed to create long-lasting memories? A few clues can be gleaned from people whose livelihood depends on their memory. Actors, for example, seek a realistic portrayal of their characters by appearing to speak lines spontaneously that they have already committed to memory.

How Actors Remember Their Lines

After a recent theater performance, I remained in the audience as the actors assembled on stage to discuss the current play and the upcoming production that they were rehearsing. Because each actor had many lines to remember, my curiosity led me to ask a question they frequently hear: "How do you learn all of those lines?" Actors face the demanding task of learning their lines with great precision, but they rarely do so by rote repetition. They did not, they said, sit down with a script and recite their lines until they knew them by heart. Repeating items over and over, called *maintenance rehearsal,* is not the most effective strategy for remembering.[3] Instead, actors engage in *elaborative rehearsal,* focusing their attention on the meaning of the material and associating it with information they already know. Actors study the script, trying to understand their character and seeing how their lines relate to that character. In describing these elaborative processes, the actors assembled that evening offered sound advice for effective remembering.

Similarly, when psychologists Helga and Tony Noice surveyed actors on how they learn their lines, they found that actors search for meaning in the script, rather than memorizing lines. The actors imagine the character in each scene, adopt the character's perspective, relate new material to the character's background, and try to match the character's mood. Script lines are carefully analyzed to understand the character's motivation. This deep understanding of a script is achieved by actors asking goal-directed questions, such as "Am I angry with her when I say this?"[4] Later, during a performance, this deep understanding provides the context for the lines to be recalled naturally, rather than recited from a memorized text. Actor Michael Caine described this process well:

> You must be able to stand there not thinking of that line. You take it off the other actor's face. Otherwise, for your next line, you're not listening and not free to respond naturally, to act spontaneously.[5]

This same process of learning and remembering lines by deep understanding enabled a septuagenarian actor to recite all 10,565 lines of Milton's epic poem, *Paradise Lost.* At the age of fifty-eight, John Basinger began studying this poem as a form of mental activity to accompany his physical activity at the gym, each time adding more lines to what he had already learned. Eight years later, he had committed the entire poem to memory, reciting it over three days. When I tested him at age seventy-four, giving him randomly drawn couplets from the poem and asking him to recite the next ten lines, his recall was nearly flawless. Yet, he did not accomplish this feat through mindless repetition. In the course of studying the poem, he came to a deep understanding of Milton. Said Basinger:

> During the incessant repetition of Milton's words, I really began to listen to them, and every now and then as the poem began to take shape in my mind, an insight would come, an understanding, a delicious possibility.[6]

In describing how they remember their lines, actors are telling us an important truth about memory—deep understanding promotes long-lasting memories.

A Memory Strategy for Everyone

Deep understanding involves focusing your attention on the underlying meaning of an item or event, and each of us can use this strategy to enhance everyday retention. In picking up an apple at the grocers, for example, you can look at its color and size, you can say its name, and you can think of its nutritional value and use in a favorite recipe. Focusing on these visual, acoustic, and conceptual aspects of the apple correspond to shallow, moderate, and deep levels of processing, and the *depth of processing* that is devoted to an item or event affects its memorability. Memory is typically enhanced when we engage in deep processing that provides meaning for an item or event, rather than shallow processing. Given a list of common nouns to read, people recall more words on a surprise memory test if they previously attended to the meaning of each word than if they focused on each word's font or sound.[7]

Deep, elaborative processing enhances understanding by relating something you are trying to learn to things you already known. Retention is enhanced because elaboration produces more meaningful associations than does shallow processing—links that can serve as potential cues for later remembering. For example, your ease of recalling the name of a specific dwarf in Walt Disney's animated film, *Snow White and the Seven Dwarfs,* depends on the cue and its associated meaning:

Try to recall the name of the dwarf that begins with the letter B.

People often have a hard time coming up with the correct name with this cue because many common names begin with the letter *B* and all of them are wrong. Try it again with a more meaningful cue:

Recall the name of the dwarf whose name is synonymous with shyness.

If you know the Disney film, this time the answer is easy. Meaningful associations help us remember, and elaborative processing produces more semantic associations than does shallow processing. This is why the meaningful cue produces the name *Bashful*.

Surviving on a Deserted Island

Demonstrating the effectiveness of meaningful, elaborative processing is easy. In class, I would read my students a list of common nouns and ask them to make a decision about each word. Half of the class had to visualize each word as I read it to determine if it contained the letter *E*, while the other half had to decide if each word represented an object they would want if they were stranded on a deserted island. Because I made no mention of any memory test, the students made their judgments with little intention to remember. Later, on a surprise recall test, I always found the same result: those who engaged in meaningful, elaborative processing—by thinking of the words as objects they might use—remembered many more words than those who engaged in letter processing. This same outcome actually occurs when people know in advance that their memory will be tested, showing that *attention*, not *intention*, is important for remembering.[8]

Deciding whether a word such as *Hammer* or *Hat* represents a useful object for your survival if stranded on a deserted island not only captures the word's meaning, it engages you in elaborative processing, making you think of potential uses for that object. Focusing attention in this way boosts retention, says psychologist James Nairne, because our episodic memory has been tuned by evolution, over our long ancestral past, to remember especially well any information—whether finding food, finding a mate, or avoiding predators—that is important for our survival as a species.[9] If, for example, you ask people to rate words representing common objects in terms of their survival value if stranded in the grasslands of a foreign land versus asking them to rate those same words for their relevance to moving into a new home in a foreign land, you will not get equivalent remembering. Survival processing yields the best memory, reflecting our brain's Stone Age past.[10]

There are numerous survival films showing individuals faced with life-threatening dilemmas, including Danny Boyle's *127 Hours*, Sean Penn's *Into the Wild*, and Chris Kentis's *Open Water*. But none depicts survival processing better than Robert Zemeckis's film, *Cast Away*, with Tom Hanks and Helen Hunt. Stranded on a deserted island with only a few man-made objects, Chuck Nolan must discover new uses for these objects, if he hopes to survive.

"I knew, somehow, that I had to stay alive."
Cast Away

Cast Away is a tale about fate—how Chuck's life was irrevocably changed by an unexpected event. It is also a lesson about faith—how his life moved in a new direction, once he summoned the courage to venture forth. As a troubleshooter for Federal Express, Chuck's job is to keep the packages moving, even if it means leaving his fiancée Kelly alone on Christmas Eve. Boarding a cargo plane bound for Malaysia and assuring her that he will be back for New Year's Eve, Chuck will not keep this date.

Flying over the South Seas, his plane is struck by a violent storm, veers badly off course, and crashes in the dark sea. Surviving the impact and blazing wreckage alone, he drifts away in an inflatable raft, before washing ashore on a tropical island. As waves lap the beach the next morning, he awakens, still dazed by what has occurred. Looking about, he cries out, "Hello? ... Anybody? ... Help! ... Help!" Slowly, the realization sinks in that he is alone on a deserted island.

Chuck's attention, now focused on survival, is directed to exploring the island, looking for food, warmth, and shelter, but his search is futile, save for a few FedEx packages that have washed up on shore with the tide. Setting aside a box with a pair of angel wings painted on it, he opens the others to find an assortment of seemingly useless items—a party dress, a pair of women's ice skates, a box of videotapes, and a Wilson volleyball. Fashioning a face on the ball and naming it "Wilson," Chuck relieves his growing loneliness through conversations with this silent companion, a cinematic device that lets film viewers in on Chuck's thoughts.

Flashing forward four years, this modern-day Robinson Crusoe has become adept at survival processing. Splitting coconuts with sharp rocks, fishing with a net made from the dress lining, using the blade of one ice skate as a knife, tying the other skate to a tree limb to form an axe, and braiding the videotapes into a rope for a raft fashioned out of tree trunks, Chuck has adapted and endured. When the side panel of a portable toilet washes ashore, it becomes the raft's sail, providing him the means to escape—if he has the courage to risk it. His choice is stark: stay on the island with the certainty of one day dying alone or risk everything by embarking on a perilous, open-sea voyage, hoping to be spotted by a ship. Choosing faith, Chuck shoves off with his volleyball and unopened FedEx package, saying, "Wilson, my main man. Time to go. Don't worry ... I'll do all the paddling. You just hang on."

Drifting for days at sea, desolate after Wilson floats away, Chuck is near death when saved by a passing freighter and returned home for a hero's welcome. At his party, a bowl of untouched crab legs serves as a poignant reminder of how hard he toiled for food on the island. Civilization requires readjustment—the biggest being his relationship with the now married Kelly. Talking with her of what might have been, tearfully accepting that their lives have gone in different directions, Kelly must stay with her family, and Chuck needs to move on. This time he is headed for Texas, hand-delivering the unopened FedEx package to the woman whose painted angel's wings guided him through the worse that fate could deliver. Who knows what the next tide will bring.

Facing the Future by Looking Backward

Survival processing requires projecting ourselves into the future, making plans that are based on memory.[11] When Chuck tore open the FedEx packages on the island, thinking how he might use each item, he engaged in *future thinking*—remembering the past to imagine the future. Relying on both memory and imagination to plan his escape, he crafted an axe from an ice skate that he used to chop trees to make a raft. Combining past experiences in novel ways, say psychologists Daniel Schacter and Donna Addis, enables us to navigate future scenarios:

> When we imagine different versions of tomorrow's big meeting ... for example, we project ourselves into the future based on what we remember from the past. Indeed, information about the past is useful only to the extent that it allows us to anticipate what may happen in the future.[12]

Navigating future scenarios based on past experience is possible because a common brain network links remembering and imagining.[13] When this neural network is impaired, people who have difficulty recalling the past—because of major depression or amnesia, for example—can have trouble imagining the future.[14]

Memory processes are meant to function both backward and forward in time. As the White Queen said to Alice in Lewis Carroll's *Through the Looking Glass*, "It's a poor sort of memory that only works backwards."

But what if there were no need for future thinking because each day was the same as the last and memory was only for remembering? Could you change your life if you were unhappy, remembering only how badly your life has turned out? This is Phil Connors's predicament in Harold Ramis's romantic comedy, *Groundhog Day*, with Bill Murray, Andie MacDowell, and Chris Elliott.

"What would you do if you were stuck in one place
and every day was exactly the same?"
Groundhog Day

Groundhog Day is a fantasy about a person caught in a time warp. Phil Connors is an irascible TV weatherman who is sent to Punxsutawney, Pennsylvania, one cold February 2nd to see if the famous groundhog will spot his shadow. According to custom, if the groundhog sees his shadow, there will be six more weeks of winter. Phil grudgingly conducts his telecast, disparaging his coworkers in the process, before retiring to the local inn for the night. But the next morning and each morning thereafter, the same day unfolds. Every morning at 6 a.m., Phil awakens to the sound of Sonny and Cher singing "I Got You, Babe," on his clock radio.

Initially baffled by the daily repetitions, Phil quickly realizes that people will repeat their actions and that he can take liberties with them. The knowledge that his actions have no lasting consequences leads him to take advantage of Rita, his charming producer, by learning of her likes and dislikes on one day, and presenting himself as a like-minded suitor the next. Duly impressed, Rita sees Phil in a new light, even though it is just a con for him. After a pleasant day together, she sighs, "You couldn't plan a day like this." Phil, knowing otherwise, replies, "You can. It just takes an awful lot of work."

Unhappy with his life, seeing no future other than repeating the same joyless day, Phil eventually comes clean to Rita, who tells him that if he wants his life to change he needs to be a better person. Taking her advice to heart, he sets out helping others, delivers an eloquent Groundhog Day newscast, and gains Rita's love and respect. The next morning, waking once more to Sonny and Cher, but this time with Rita beside him, Phil realizes that the time warp is broken.

> Phil: Something is different ... You know what today is?
> Rita: No. What?
> Phil: Today is tomorrow. It happened.

This allegory offers a lesson. Whereas *After Life* shows that memory provides the opportunity for reflection, *Groundhog Day* offers us hope, showing how memory allows us to change. Unlike the characters in *After Life* who happily relive their favorite memory forever, Phil in *Groundhog Day* keeps repeating the same dreadful day, until he realizes that only he can be the agent for change in his life.

The feeling that all of us occasionally experience of having seen or heard something before, but knowing that we have not, is called *déjà vu*.[15] Phil in *Groundhog Day* is not experiencing déjà vu; he knows that he keeps reliving the same day. Yet there are individuals with brain damage in their medial temporal lobes that sometime have the persistent feeling of having lived the present moment before, a phenomenon called *déjà vécu*. One elderly man, for example, refused to read the day's newspaper or watch television, claiming he had already done so. When asked by his wife what happened next in a program he claimed to have seen, his reply was insightfully evasive: "How should I know," he said, "I have a memory problem!"[16] Déjà vu is a relatively common experience owing to a sense of familiarity when we are subtly reminded of something from our past; déjà vécu is uncommon and a sign of brain dysfunction. Phil's experience of reliving the same day is pure fiction, but as a metaphor for getting out of a rut, it illustrates the power of future thinking.

Different Ways of Remembering

Can we remember nonconsciously? When people are repeatedly flashed a series of random shapes, one after the other, but too briefly to be clearly seen, they will have no conscious memory of those shapes. When later given pairs of shapes on a test, each containing one old and one new, people will be unable to tell which shape they were previously shown. However, as my students and I have demonstrated, if asked to select the shape they like best in each pair, people often pick the previously shown shape. This preference, called the *mere exposure effect,* is a nonconscious form of remembering.[17] Psychologists have long been tantalized by the idea of nonconscious mental processes—thoughts we cannot think, emotions we cannot express, and memories we cannot recall. These processes, long elusive to scientific study, have been uncovered during the past several decades, revealing how we can be influenced by past experiences without our awareness. Remembering, we now know, can be implicit as well as explicit.

Remembering With and Without Awareness

The presence or absence of awareness distinguishes explicit and implicit remembering. *Explicit remembering* involves the conscious recollection of a prior experience—as in, for example, your awareness of mentally traveling backward in time to recall what you did last weekend. *Implicit remembering* lacks this awareness. It entails changes in your thoughts or actions produced by a prior experience without any conscious recollection. In making a new acquaintance at a party, for instance, you might take a sudden liking or disliking to this person because she or he reminds you of someone from your past, but you are unaware of this association—your past has implicitly reached into the present. One type of implicit remembering that we have all experienced is *intuition*, a sense of knowing or understanding that we feel without conscious recall.

Neurological case studies reveal that these two types of remembering are independent. One patient with brain damage in his occipital and temporal lobes could not consciously remember previously studied faces, a disorder called *prosopagnosia*. Yet, when he was later shown old faces paired with new faces and asked to select those he liked best, he demonstrated a mere exposure effect, selecting faces he had previously seen. He remembered faces implicitly that he was unable to recognize explicitly.[18] Another patient with a damaged occipital lobe demonstrated explicit remembering without implicit remembering, suggesting that these two forms of remembering are governed by different neural systems in the brain.[19]

Sometimes recent experiences can implicitly nudge us in surprisingly subtle ways. Take, for example, the concept of *warmth*. We can think of warmth physically as a measure of heat or psychologically as a personality attribute. When psychologists Lawrence Williams and John Bargh had people hold a cup of hot or ice coffee while rating the psychological warmth of a person described with positive traits, people rated that person as having a warmer personality if they previously held the warm drink.[20] Similarly for *closeness*: when people connected two points on a graph that were physically close together, they subsequently rated their family relations as emotionally closer than people who connected more distant graph points.[21] In each instance, a concept's physical dimension—warmth or closeness—implicitly influenced its psychological dimension. Holding a cup of coffee or drawing a line on a graph should have no bearing on your social judgments, but they do because these concepts, once aroused in the brain, can generalize, exerting their influence without your awareness.

Prejudicial Stereotypes

Stereotypes can also influence us implicitly. A *stereotype* is a set of positive or negative beliefs ascribed to a social group. Negative racial stereotypes, for example, have been demonstrated even in people who explicitly reject such prejudicial beliefs. When white college students were presented with a list of words that were stereotypically associated with black Americans—words such as *ghetto*, *busing*, and *Harlem*—they later rated a man of unspecified race as more hostile than students who were exposed to neutral words, even though the words were presented too rapidly to be clearly seen.[22]

Gender and cultural stereotypes may similarly influence our behavior without awareness. Asian-American women who completed a questionnaire about female identity scored lower on a subsequent math test than those who previously answered questions related to their cultural identity. For math, there are conflicting stereotypes: a *negative* gender stereotype that women are poor at math and a *positive* cultural stereotype that Asians are good at math. Merely answering questions pertaining to either their gender or cultural background was enough to activate a stereotype implicitly that biased their math test result.[23] Formed early in life, implicit prejudicial attitudes can linger long after explicit attitudes have become more egalitarian.[24] Paul Haggis's film, *Crash*, with an ensemble cast including Sandra Bullock, Don Cheadle, Matt Dillon, Terrence Howard, and Chris Bridges, shows how social stereotyping can lead to dangerous consequences.

"Think we miss that touch so much, we crash into each other just to feel something."
Crash

Prejudicial stereotypes propel the characters in *Crash*—sometimes knowingly, other times without their awareness—into multiple physical and psychological collisions. Depicting events over two days in Los Angeles, *Crash* tells interlocking stories of characters who differ in race, income level, and culture. It is a film about in-groups and out-groups—blacks and whites, cops and criminals, rich and poor, natives and immigrants—and what happens when their prejudices collide.

Early on we meet Anthony and Peter, two young black men, dressed as college students, exiting a restaurant in a white neighborhood. Upset by the restaurant's service, Anthony utters a negative stereotype, saying, "That waitress sized us up in two seconds. We're black and black people don't tip, so she wasn't gonna waste her time." When a white couple approach on the sidewalk and the woman subtly takes her husband's arm for protection, Anthony is again quick to notice:

Anthony: You see what that woman just did?
Peter: She's cold, man.

> Anthony: She got colder soon as she saw us ... Look around! You couldn't find a whiter, safer or better lit part of the city. But this white woman sees two black guys ... her reaction is blind fear.

Her fear is quickly realized when Anthony and Peter hijack their car.

Across town, after failing to get help for his ailing father, police officer John Ryan insults a medical supervisor named Shaniqua Johnson, saying that she got her job through a hiring quota. Later, while on patrol, he pulls over a vehicle driven by a black couple that resembles the vehicle stolen by Anthony and Peter. Even though the license plate is not a match, Ryan humiliates the woman with a full body search, forcing her powerless husband to watch. Embarrassed by this explicit racism, Ryan's rookie partner, Tom Hanson, requests a reassignment. Wishing Hanson well, Ryan's parting words will be eerily prophetic: "Wait 'til you've been doing it a few more years. You think you know who you are? You have no idea."

Later that night, Hanson offers a cold hitchhiker a lift; it turns out to be Peter trying to get home. Making small talk in the car, Peter describes his enjoyment of country music and his love of ice hockey—talk that conflicts with Hanson's racial stereotypes about blacks. Amused by Hanson's confusion and spotting a St. Christopher figure on Hanson's dashboard—the same statuette that he carries in his pocket—Peter starts laughing. But Hanson, misinterpreting his laughter as ridicule, orders him out of the car as Peter reaches into his pocket to explain. But Hanson's racial stereotypes are now aroused. Believing that Peter is reaching for a gun, Hanson pulls out his revolver and shoots him. It is over in an instant. Explicitly rejecting racial prejudice, Hanson acted on a stereotype that was unconsciously triggered by Peter reaching into his pocket, an act that biased Hanson's judgment in the heat of the moment. *Crash* is full of these powerful examples of people acting on stereotypes—beliefs about others harbored away in memory.

Where Do We Keep Our Memories?

Which statement comes closest to your belief about memory?

A. Everything we learn is permanently stored in the mind, although particular details are not always accessible.

B. Some details that we learn may be permanently lost from memory.

When given this choice, many people selected Statement A, even though it is without scientific support. When asked to explain, some people cited a personal experience of recalling an event that they had not thought of for some time, while others said that they heard of the brain operations performed in the 1950s by Canadian neurosurgeon Wilder Penfield.[25] Penfield's operations are famous, but they are not evidence of memory permanence.

In aiding epileptic patients whose convulsions were not responsive to drugs, Penfield pioneered a procedure for removing the brain area that generated the seizures. Awakened during the operation, as Penfield applied electrical stimulation to map the brain's surface, patients occasionally reported what Penfield thought to be memories, leading him to conclude that the brain kept a permanent record of a person's experiences.[26] However, an analysis of the hundreds of patients who received this procedure revealed that these reports occurred less than 8 percent of the time and were often more like dreams than memories. Moreover, surgically removing specific brain regions to control epilepsy did not erase the memories supposedly associated with those areas.[27] Memories are not stored in specific brain locations and, even though all of us occasionally recall events we had not thought of for some time, we cannot assume that all memories are permanent.

Peering inside the brain using modern imaging techniques such as fMRI and PET scans reveals that multiple brain regions are involved in various memory tasks. The brain is still the place our memories call home, but memory is now thought of as an emergent property that arises from the activity of numerous neural networks that are distributed over the brain.[28] When we experience an event, it is broken down into its various components and processed in different brain regions to give our memories a multisensory quality including sights, sounds, and emotional coloring. A memory is the brain's record of an event—a connection among these distributed neural networks that fired together during an experience and fire again when that experience is recalled.

We remember many things from our past, including experiences not thought of for years. But that does not mean that every single past experience, the big and the small, is tucked away inside the brain, just waiting for the right cue to pop forth. Some experiences are forgotten, and this serves an adaptive function by eliminating useless clutter. It is part of what the brain does. On occasion, we might like to guide this process, selectively eliminating those memories of people or experiences we no longer want. But this type of targeted memory erasure, shown in Michel Gondry's *Eternal Sunshine of the Spotless Mind,* featuring Jim Carrey, Kate Winslet, and Tom Wilkinson, exists only in science fiction.

"Please let me keep this memory, just this one."

Eternal Sunshine of the Spotless Mind

After a romantic relationship ends, are the memories of that relationship worth keeping? Taking its title from Alexander Pope's poem about a tragic love affair, *Eternal Sunshine of the Spotless Mind* tells the strange story of a lonely guy named Joel who falls in love with a quirky bookstore clerk named Clementine. Her color-streaked hair tells us where we are in their relationship: blue for the present, red during their prior two years together, and green for their first meeting—a helpful guide because much of this film takes place in Joel's mind. Jumping back and forth in time, the story begins with Joel skipping work one February day, heading to Montauk, where he sees Clem walking alone on the beach. Too shy to make eye contact with her, he lets the opportunity pass until she eventually strikes up a conversation, asking on their train ride home, "Do I know you?" In fact, both do, but neither is aware that they were former lovers who met in Montauk before.

After a bitter argument, Clem ended their earlier relationship and eliminated her memories of Joel by going to Lacuna Incorporated, a local memory clinic that erases the memories of former lovers but leaves other memories intact. Wounded by Clem's action, Joel retaliates by visiting the clinic, wanting Clem similarly erased from his mind. Dr. Mierzwiak, the inventor of the procedure, tells Joel to bring in all of the objects that remind him of Clem. By scanning Joel's brain while he looks at the various objects, Mierzwiak targets Joel's memories of Clem. Afraid, yet determined to proceed, Joel questions Mierzwiak about the procedure:

> Joel: Is there any risk of brain damage?
> Mierzwiak: Well, technically speaking, the operation is brain damage, but it's on a par with a night of heavy drinking. Nothing you'll miss.

That evening, the procedure is applied. While Joel sleeps, his memories of Clem are located and, one by one, these fragmented memories dissolve. But his erasure is incomplete. Still in love with Clem, Joel successfully fights the procedure, hiding a single memory, one of her whispering to him—"Meet me in Montauk"—where our story opens and ends.

The targeted memory erasure shown in *Eternal Sunshine* is based on the idea that specific memories can be precisely located in the brain and destroyed. While it is true that people who suffer a cerebral concussion or receive electro-convulsive shock treatment can experience temporary or permanent amnesia, their personal memory loss is never so specifically defined as this movie suggests. People can lose large chunks of time from their past, but they never lose, say, only those times associated with a former lover. If targeted erasure were actually possible, erasing the unwanted memories of a specific person would require accessing and destroying every neural representation of that person—meaning every object, place, smell, and thought associated with that person. Because these distributed neural representations necessarily link many different brain areas, the result would be massive memory loss due to massive brain damage—not at all "on a par with a night of heavy drinking." Joel would forget Clem and much, much more. A spotless mind might seem desirable, but it is more adaptive to remember and learn from our failed relationships than to forget and repeat them. Some people, in fact, are exceptionally good at remembering.

Memory Champs, Mnemonics, and Savants

Remembering appears easy for contestants in the yearly World Memory Championship where competitors from all over the world try to retain as much information as possible in a limited period of time. People are fascinated with others who can remember vast amounts of seemingly trivial information. These people, called *memorists*, can learn and remember large quantities of information quickly, yet their superior memory is not based on exceptional intelligence.[29] Exceptional memorizers, says psychologist Anders Ericsson, are made, not born.[30] Writer Joshua Foer, for example, studied for a year to reach the finals of the US Memory Championship in 2006 by successfully memorizing long lists of numbers and words, historical facts and dates, and even the order of two shuffled decks of playing cards.[31] He achieved this goal by using ancient mnemonic techniques.

A Memory Lesson from the Ancients

Mnemosyne was the personification of memory in Greek mythology and the origin of the term *mnemonic*, a device for assisting memory. To see the power of a mnemonic, study the following ten rhymes until you can recite them without looking. Trust me, they are easy to learn, and this demonstration will surprise you.

One is a bun. Two is a shoe. Three is a tree. Four is a door. Five is a hive. Six are sticks. Seven is heaven. Eight is a gate. Nine is a line. Ten is a hen.

Now that you have committed these rhymes to memory, we will use them as a scaffold for remembering these ten words:

Ghost, Banana, Elephant, Juggler, Monkey, Drum, Cannon, Tomato, Clown, Boat

Take each rhyme and word in pairs, starting with the first rhyme and first word, and form a mental picture connecting the rhyme and word. For example, for *One is a bun* and *Ghost*, you might imagine a ghost taking a bite out of a large bun. Once you have your first rhyme-word image, go on to the next pair and make a second mental picture connecting those two items. Your mental pictures are your own creations. Just be sure to connect the word and rhyme together in your mental image. After doing this for each pair, resume reading, without going back to the mnemonic. We will return to it in a bit.

This mnemonic is a simplified version of the *method of loci*, the method of locations. It is one of the oldest mnemonic devices, dating back over 2,000 years to ancient Greece. In using this device, you would typically use a set of familiar locations, such as the rooms in your home or shops in your town. To remember a long list of items, you would take a mental walk through your home, visualizing the objects already in each room, and imagining each item you wish to remember connected to a different object along the way. To remember later, you take a mental walk back through your rooms, picking up each item that you left with each object. Like the rhyme scheme, the method of loci works because it provides a plan for remembering: it tells you where to start, how to proceed, and when you are done. Now, starting with *One is a bun*, recall all ten words in order by reciting the rhymes, recalling your mental pictures, and saying the words. This task should be surprisingly easy.

Forming mental images of the items that you wish to remember and connecting them to form vivid, imaginary stories provides the basis for many powerful mnemonic demonstrations. Rajan Mahadevan was able to memorize railway timetables and large arrays of random digits; Chao Lu memorized pi, the mathematical constant, to 67,890 decimal places in this manner.[32] Each man spent thousands of hours honing his mnemonic technique so that he could quickly turn long strings of digits into meaningful mental pictures, weaving them together into an imaginary story for later recall. Both combined elaborative processing with a plan for remembering, just as you did on a much smaller scale with the rhyming mnemonic. Some people, however, show extraordinary skill in remembering that is not based on using mnemonics.

The Puzzling Memory of Savants

A quick online search for Daniel Trammet, Derek Paravicini, or Stephen Wiltshire will provide mind-boggling memory demonstrations that do not seem humanly possible. Trammet learned the Icelandic language from scratch in one week. Paravicini is mentally disabled, yet plays a repertoire of thousands of songs flawlessly on the piano. Wiltshire, also mentally disabled, draws aerial views of Rome and Tokyo with every building and landmark in place after a helicopter ride over each city.

These men demonstrate *savant syndrome* by exhibiting a profound mental ability often accompanied by low intellectual function or *autism,* a disorder involving impaired social interaction, communication problems, and repetitive actions. While mentally impaired, they are capable of brilliance in art, music, or math as a result of their prodigious memory. This dazzling yet mysterious ability is explored in Barry Levinson's *Rain Man,* with Dustin Hoffman, Tom Cruise, and Valeria Golino. Inspired by the real-life autistic savant, Kim Peek, who began memorizing books at the age of eighteen months, *Rain Man* shows the benefits and costs of having a memory that psychologist Darold Treffert describes as deep, but exceedingly narrow.[33]

> "Why didn't anyone tell me that I had a brother?"
>
> *Rain Man*

Rain Man is a heart-rending story of two brothers who, separated during early childhood, discover each other during adulthood.

Raymond Babbitt, a mentally disabled autistic savant who lives each day by routine, was institutionalized after his mother's death. His brother Charlie, a callous, fast-talking car salesman, grew up unaware of Raymond's existence.

Told of his estranged father's death, Charlie flies home to Cincinnati for a reading of the will and learns that he has inherited his father's 1949 Buick convertible, along with his prized rose bushes. But the money from the estate, some three million dollars, will be handled by Dr. Bruner, a trustee who is charged with caring for Raymond. Discovering that he has a brother and feeling cheated over the will, Charlie packs Raymond in the Buick and sets off for Los Angeles, planning to wrest half of his father's estate from Bruner. What follows is a road film in which Charlie learns the meaning of compassion by taking the road less traveled.

Self-centered and needing to rush back to Los Angeles, Charlie soon learns that Raymond has nonnegotiable needs of his own. Citing the accident records of airlines, Raymond will not fly on a plane—causing Charlie to drive across country; nor he does like highways—causing Charlie to take the back roads. Even a pair of underwear generates chaos when Raymond informs Charlie that he is not wearing any, saying, "They're not my underwear ... Mine are boxer shorts." Charlie, failing to understand, exclaims, "What's the difference? ... Underwear is underwear." For Raymond, changes are frightening; safety is found in routines. For him, lunch must be at 12:30, he must watch *People's Court* and *Wheel of Fortune*, his bed must be by a window, he must eat pancakes for breakfast with a toothpick, and lights must be turned off at 11 p.m.

Yet, Raymond possesses astonishing memory and math skills. During their cross country odyssey, he memorizes all of the names and phone numbers of people from A though G in a town directory, accurately counts the 246 toothpicks dropped on the floor, solves complex math problems such as finding the square root of 2,130 in his head, and counts cards in Las Vegas to win at blackjack from six decks of cards. Asked by Charlie how he does this, Raymond says that he just sees the answers. Yet, simple subtraction is beyond his grasp. When he is asked, "If you had a dollar, and you spent fifty cents, how much money would you have left?" Raymond replies, "Seventy cents." This wiz at mental calculation is paradoxically mentally impaired.

A turning point in the film occurs one evening when Raymond, while brushing his teeth, repeats a conversation he had long ago with Charlie:

> Raymond: You said funny teeth, funny Rain Man.
> Charlie: Rain man? I said, "Rain man"?
> Raymond: Yeah, funny Rain Man.
> Charlie: Was I trying to say "Raymond" and it came out "rain man"? ... You? You're the rain man?

Growing up believing that the Rain Man was an imaginary childhood friend, Charlie finally sees Raymond as family—the older brother who sang him Beatles' songs. Finally accepting that Raymond is better off with Bruner and declining an offer of money, Charlie brings his brother to the train station, saying, "I'm coming to see you in two weeks." Preoccupied with his portable TV, Raymond nods, adding that will be in 14 days or 336 hours. "Mystifying," is all Charlie can say. The brother incapable of change has taught Charlie a lesson in compassion. Definitely.

Remembering in the Internet Age

The memory abilities of real-life savants are poorly understood, but they do not have *photographic memory*, a fictional ability to recall experiences in perfect detail without the aid of mnemonics. People sometimes claim to know someone who possesses this mythical ability, yet no reliable evidence exists. The closest related real phenomena are *eidetic imagery* and *hyperthymesia*. Eidetic imagery is a form of visual memory occasionally found in grammar-school children. These youngsters can continue to see and describe a visual scene after it has been removed, but only for a matter of seconds, and they lose this ability as they age.[34] Hyperthymesia refers to the exaggerated remembering of personal and public events. One woman with this condition could accurately recall what she did each day over many years. Yet, by continually dwelling on the past, she saw herself as a prisoner of her memories, unable to lead a normal life.[35]

Mentally disabled savants demonstrate exceptional memory in specific domains, but they, too, are unable to lead normal lives. We might marvel at people like Kim Peek, but none of us would willingly trade places. *Rain Man* dazzles us with demonstrations of prodigious long-term memory, while also showing how futile life would be without the ability to use memory for future thinking. Raymond can memorize much, but he is unable to put it to use on his own. Each of us armed with a laptop has access to more information than Raymond or any real-life savant could acquire in a lifetime. Still, as Charlie said, their memory feats are mystifying.

Today, our laptops, tablets, and smartphones are changing how we use memory, making us less dependent on memorization and more reliant on knowing how to access and evaluate information. If we need to know the latest news, weather forecast, or film review, the information is just a few clicks away. With constant access to information online, our need to remember it is reduced, while our need to remember where to find it is enhanced. As psychologist Betsy Sparrow and her colleagues have found, people will forget information that they believe is readily available online, but remember information that they think will be unavailable.[36] Smart technology changes what we remember, but it will not make memory obsolete. By combining information on the Internet with our personal experiences, memory becomes more powerful, helping us better plan for the future.

Fade-Out

Strategies involving deep, elaborative processing produce long-lasting memories that enable us to imagine and plan. Without future thinking, even the prodigious memories of savants serve no useful function. Mnemonics facilitate remembering, but their everyday use is of limited value, diminished more so by the vast amount of information available on the Internet. Smart technology can enhance future thinking by making us less dependent on memorization and more reliant on locating and evaluating information.

4 Recognizing the People We Know

Setting the Scene

What is the essence of a person that permits recognition? Is it his or her appearance, personality, shared memories? Which characteristics are more reliable than others, and what makes them useful for identification? The films in this chapter play on the theme of person recognition as each involves a potential impostor in the family. In examining these movies, think of the various cues that family members employ to distinguish imposters from loved ones. Physical features need not be the most reliable cues for identifying the people we know.

Featured Films

The Imposter, docudrama directed by Bart Layton, with Frédéric Bourdin (Picturehouse Entertainment, 2012, Rated R).

The Return of Martin Guerre, directed by Daniel Vigne, with Gérard Depardieu and Nathalie Baye (European International, 1982, French with English subtitles, Unrated).

Face/Off, directed by John Woo, with John Travolta and Nicholas Cage (Paramount Pictures, 1997, Rated R).

Invasion of the Body Snatchers, directed by Philip Kaufman, with Donald Sutherland and Brooke Adams (United Artists, 1978, Rated PG).

Changeling, directed by Clint Eastwood, with Angelina Jolie and John Malkovich (Universal Pictures, 2008, Rated R).

Seeing With Our Heart, Not Our Eyes

Intuitively, we think that distinguishing an imposter from a family member would be easy because we know our loved ones so well. Yet sometimes this belief is surprisingly wrong. As shown in Bart Layton's dramatized documentary, *The Imposter,* families have made some stunning recognition errors in welcoming a lost child home.

"No one would be wrong about something like that."
The Imposter

The Imposter tells the story of Frédéric Bourdin, a then twenty-three-year-old Frenchman—nicknamed *the chameleon*—who had a history of impersonating lost children.[1] Slight of build, Bourdin searched databases to find missing children, then altered his appearance to suit his current needs. One search turned up Nicholas Barclay, a sixteen-year-old boy with blond hair and blue eyes who went missing from his Texas home three years earlier. Bourdin was not only seven years older than Barclay, his hair and eyes were brown, and he spoke English with a foreign accent. Undeterred, Bourdin gained information about this case on the Internet, including the names and locations of family members, before turning himself over to Spanish authorities, telling a bizarre tale of having been kidnapped and tortured by European child traffickers.

When Bourdin was delivered to the Barclay family, he came with dyed hair and an elaborate, concocted story to explain his physical differences from the real Nicholas. Surprisingly, the family accepted him, even as he mined them for information about past events, later discussing those events with others to show that he remembered family life. Any doubts about his identity were attributed to his long ordeal. Said Carey, Nicholas's sister, "Your heart takes over and you just want to believe."

Upon news of Nicholas's return, a private investigator grew suspicious and uncovered a number of discrepancies—photos of Nicholas and Bourdin showed that their ears did not match, an ophthalmologist disputed Bourdin's claim that his abductors could have chemically changed his eye color, and a language expert noted that the real Nicholas would not speak English with a foreign accent. Yet Nicholas's mother angrily dismissed this evidence. Only after an FBI agent got a warrant to obtain Bourdin's fingerprints and a sample of his blood did these efforts expose this cruel hoax. When asked how she could have accepted an older man with dyed hair, dark eyes, and a foreign accent as her son for five months, the mother replied, "We just kept making excuses, that he's different because of all this ugly stuff that had happened." Sometimes we see what we want.

Despite glaring discrepancies, this daring con artist played on a family's longing to gain temporary acceptance as their lost son. Eventually, the essential elements of recognition could no longer be denied, revealing what was obvious all along.

Recognition's Essential Elements

Physical features such as a person's face and voice play a critical role in recognizing others, but so do less discernible characteristics such as someone's personality and mannerisms. For our family and friends, we know what they *look* like, and we know what they *are* like. Consider, for example, what it would take for an imposter like Bourdin to fool us by impersonating someone we once knew well, but had not seen in a number of years. We all undergo change, but three relatively stable characteristics play essential roles in recognizing others. The first involves a person's physical features, including their appearance and voice; the second entails someone's psychological characteristics—their personality, mannerisms, and peculiarities; and the third includes shared memories, events that you experienced with another person that permits shared reminiscence. In short, we recognize the people we know by how they look and sound, how they act, and what they know.

Physical Features

The face is our most varied attribute, and while it is tempting to think that our facial features remain constant, they actually undergo slow but continual change. Underlying the structure of the face are fourteen bones that differ in size and shape, covered with a layer of fatty tissue that varies in smoothness and thickness in each of us. At birth, our forehead tends to be large and our cheeks shallow, but changes occur as we age. We lose our high forehead, the distance between the level of our eyes and nose grows greater, and our once small, concave nose grows broader and flatter. With advancing years, our eyes become less lustrous and our skin less elastic. Gravity, the ever-diligent agent for change, tugs our skin down over time, causing sags and wrinkles to begin in our twenties that become noticeable by our forties.[2] Yet, for those we knew during childhood, we may still see the child in the adult.

Our voice is another defining feature, evidenced by a recent anecdote. On a whim, I called a former high school friend, someone I had not spoken to in many years. When my friend answered the phone, I asked, "Is this Mal?" whereupon he yelled out my name. Still later, he remarked that I still laughed the same. For the people we know, we remember multiple identifying characteristics, even though we might not have noticed ourselves paying attention to these features at the time.

Psychological Characteristics

Describing a person psychologically involves identifying his or her *personality traits*: those enduring patterns of behavior showing how someone typically perceives, relates to, and thinks about him- or herself in personal and social situations. Five major traits are essential for describing someone's psychological characteristics:

Neuroticism: Is the person anxious and worrisome or calm and composed?

Extraversion: Is the person confident in interpersonal relations or shy and retiring?

Openness: Does the person try new experiences or stick with the tried and true?

Agreeableness: Is the person warm and friendly or cold and quarrelsome?

Conscientiousness: Is the person reliable or irresponsible?

These traits are measured by personality tests, consisting of a series of opposing adjectives. For example, for the trait of neuroticism, on a scale of 1 to 7, where would you place yourself along this dimension if *Calm* is a 1 and *Anxious* is a 7? You may score high or low on this dimension or anywhere in between.

Answering these types of questions for each of the five traits yields a personality profile, a psychological description of what you are like as a person. These traits make their appearance in childhood and become relatively stable by early adulthood, and this stability is beneficial, say psychologists Robert McCrae and Paul Costa, because it provides each of us with a measure of predictability—meaning that we can make choices about jobs, friends, and life partners knowing that our interests and values will not quickly abandon us.[3]

Personality stability also provides an important clue for person recognition. For the people we know, we supplement our knowledge of their physical features with knowledge of their psychological traits, creating, in effect, *physical portraits* of these people that show what they look like and *psychological portraits* that reveal how they typically behave.

Shared Memories

Much of the joy in renewing old acquaintances is the opportunity to reminisce over shared experiences, recognition's last essential element. Our memories will not always match those of an acquaintance, but for many such events—whether a summer vacation or a family wedding—we expect the memory of others who shared those earlier times to be generally similar to ours. Shared memories are part of how we come to know others and how they come to know us; in losing them, we lose a defining feature.

The impact of losing shared memories will be evident in a later chapter when we consider the terrible consequences of dementia. In the films *Amour* and *Away from Her*, the wife in a long-term marriage no longer remembers her husband and important events from their past. Although her appearance is unchanged, her mental deterioration has made her a stranger, robbing her of an essential aspect of her identity—the memories she shared with her husband, memories that helped to define their marriage. For the people we know, we expect them to remember the events that we experienced together, more or less in the same manner as we do.

Remembering shared experiences is important, but it is not sufficient for recognition. In David Vigne's *The Return of Martin Guerre*, Martin's wife Bertrande clearly remembers her long-absent husband and the early years of their marriage. But when a man claiming to be her husband suddenly returns, recalling many intimate details of their marriage, she will need all three recognition essentials—physical features, psychological characteristics, and shared memories—to make an accurate identification.

"Bertrande, your husband's back!"
The Return of Martin Guerre

Based on an actual courtroom trial that took place in rural sixteenth-century France, *The Return of Martin Guerre,* featuring Gérard Depardieu and Nathalie Baye, provides an engrossing lesson in person recognition. This historically detailed film tells the strange story of a fourteen-year-old farm boy named Martin Guerre who married Bertrande de Rols, the daughter of a local, well-to-do family. Their marriage produced one son, but after eight years, Martin wandered off to the wars in Spain where he stayed for over nine years. Later we observe a thirty-two-year-old man claiming to be Martin returning home to his family and now prosperous farm.

Martin's return is initially a cause for great celebration as he engages with his family and former friends. This man resembles the once younger Martin in appearance, and he remembers numerous prior events, including many personal details shared only with his wife. But after a few years have passed, some relatives grow suspicious, wondering if he really is Martin. Charges are brought against him, and the villagers hotly debate this issue, some testifying that this man is obviously Martin, others disputing their claim. Even Bertrande testifies in court that he is her husband. Yet doubt persists. Is this really Martin Guerre, or is he a clever imposter who resembles the real Martin and knows much of his past?

Away from home for a lengthy time, maturation would have changed Martin's appearance, yet his personality should be largely unaltered, and he should remember important events from his past. In observing Martin before he left home and after his return, film viewers get to examine his physical features—including his appearance and voice, his personality traits—such as his level of extraversion and agreeableness, and the accuracy of his memory—events shared with family and friends. As the story unfolds, Bertrande uses all of these cues to determine if this man is her husband or a shrewd imposter. Recognizing people we once knew is not always easy, and sometimes their psychological characteristics can be just as important as their physical features.

> Martin's sister: When you left here, Martin, you didn't have a beard yet, and you didn't drink as much.
> Martin: Now I drink like a monk.

The Return of Martin Guerre is important for understanding person recognition because it questions the simplistic, implicit assumption that recognizing others is a trivial matter, based solely on a person's appearance. Recognition occasionally fails us, sometimes in surprising ways.

How Good Are We at Seeing Others?

A few years ago I experienced a startling event. On the first day of classes, I met with a group of thirty students for more than an hour, and nothing unusual happened. The shock came later that evening when my wife and I dined at a local restaurant. As we began eating, three students entered and sat at a table close by. Everyone made eye contact, and then quickly resumed their conversations. The students were discussing their new courses, including one student, sitting diagonal from me, who earlier that day had sat in my class and who now discussed my course with her friends. In hearing her comments, I glanced over and was tempted to say, "Hey, don't you recognize me? I am right here." But the psychologist in me just smiled, content to let this scene play out. How long would it take her to realize that I was sitting nearby in plain sight?

Surprisingly, she never realized that I was at the next table until I discreetly described the incident when we met again, and she embarrassingly said that she'd had no idea I was there. For all intents and purposes, I was invisible to her that evening. She got a good, long look at me in class and still failed to recognize me a few hours later. As counterintuitive as this anecdote seems, there are sound reasons why her inability to see me at dinner may not be so implausible.

The Invisible Gorilla

Psychologists Daniel Simons and Christopher Chabris conducted an ingenious study. They had people watched a video showing two teams tossing basketballs. One team, dressed in black, passed the ball back and forth to each other, while another team, dressed in white, did the same with another ball. Viewers were instructed to focus on one team and ignore the other by counting the number of times their team passed the ball. Halfway through the video, a person in a dark gorilla suit walked across the screen, right through both of the teams. Later, the viewers were asked if they noticed anything unusual in the video. If they failed to mention the gorilla, they were specifically asked, "Did you see a gorilla walk across the screen?"

Most of the people attending to the team in black noticed the gorilla because their attention was focused on dark colored figures, but, incredibly, roughly half of the people attending to the team in white missed the gorilla entirely because they had tuned out the dark figures. For them, the gorilla was hidden in plain sight. This phenomenon, called *inattention blindness*, shows that we can be blind to events in our world when our attention is focused elsewhere.[4] In this light, the failure of my student to recognize me at dinner is understandable. By her focused attention on her friends, I became the invisible gorilla.

The Unseen Map-Holder

We can also be blind to seemingly obvious changes around us. Imagine, for example, giving directions to a stranger, and without your awareness, a switch is made, and you are now giving directions to a different person. Intuitively, we think that we would notice this blatant change. But once again, our intuition is incorrect. In this scenario, a man with a map approaches a pedestrian and asks for directions. While these people are face-to-face, two men carrying a door walk between the map-holder and pedestrian. Covertly, the man with the map changes places with one of the door carriers so that after the door passes, there is a new man with a map facing the pedestrian. Many people, sometimes half or more, demonstrate *change blindness* by failing to realize that they are talking to a different person.[5] We are susceptible to this type of blindness whenever we broadly categorize other people by their age, race, or occupation, rather than focusing on the unique individual standing before us. Said one person who was fooled, "Oh, I just saw him as a construction worker. I didn't see him as an individual person." Looking is not the same as seeing.

An extreme example of blindness to change is shown in John Woo's action film *Face/Off,* featuring Nicholas Cage and John Travolta as a criminal and FBI agent who swap faces and inhabit each other's world. Family members, initially blind to this switch because they rely on appearance for recognition, gradually realize that each man's physical features are not as trustworthy as their personality and shared memories.

> "I don't know, Sean, you just seem so different."
>
> *Face/Off*

In *Face/Off,* two dangerous terrorists, Castor and Pollux Troy, threaten to destroy Los Angeles by hiding a biological bomb in the city, and FBI agent Sean Archer must stop them. In the first of many gun battles, Sean and his fellow agents capture the evil brothers. Pollux is thrown in jail, while Castor is secured in a medical facility where he is going to give up his face. In a bizarre plan to find the bomb, doctors provide Sean with Castor's identity, courtesy of a high-tech face transplant and speech-alteration software.

As Castor's physical double, Sean is jailed with Pollux, hoping to discover the bomb's location. But as he quickly learns, appearance is not enough. To be accepted as Castor, Sean must reminisce with Pollux and act tough around other convicts. The plan goes awry when the real Castor escapes, forcing the doctors to give him Sean's face and voice, while destroying all evidence of the switch. With Sean securely locked in jail and Castor on the loose, both imposters have swapped lives. Castor is now an FBI agent with a wife and daughter at home, and Sean is behind bars with a far-fetched story about a face change that no one believes.

What does it take for people to catch on to this unthinkable switch? The same recognition elements that were important for Bertrande in *The Return of Martin Guerre* apply here. Sean and Castor have the right stuff physically, but not psychologically. The kindly Sean must repeatedly remind himself to act tough, whereas the evil Castor is strangely amorous around Sean's surprised wife and overly permissive with his teenage daughter. People are perplexed by the puzzling changes they observe in these men, never suspecting that they are not who they appear to be. It is only after Sean escapes from prison and confronts his unbelieving wife with evidence about his blood type and a shared memory of their first date does she accept that a switch has been made. Speaking in Castor's voice, Sean says,

> I remember I once took a date out for surf and turf, not knowing that she was a vegetarian. So, she ate bread and broke her tooth on a rye seed. … When I finally brought her home, even though it must have hurt like hell—you kissed me.

In time, people come to see the real Sean and Castor behind their swapped facades, demonstrating the importance of personality and shared memories for recognition. Still, in showing how both men were initially accepted solely on the basis of their physical characteristics, *Face/Off* illustrates how this element is foremost in person identification.

How Do We Recognize a Face?

Normally, our ability to recognize the people we know functions so rapidly and automatically that it is difficult to appreciate the complex mental operations involved in this process. To begin understanding these operations, consider why faces are so interesting to look at. All of our social behavior involves recognizing qualities in others, and our faces provide a window to our thoughts. If we silently move our eyes and mouth, forming a grin or a grimace, others quickly grasp these universally understood expressions.[6] Our faces tell people nonverbally about our mood, intention, and attentiveness, while simultaneously providing a vital evolutionary function: our faces usually provide the most reliable means of identifying us, a function that was exploited by the characters in *Face/Off*. How do we recognize the people we know? Psychologists Vicki Bruce and Andy Young have sketched an outline of this recognition process.[7]

Seeing a Face in Stages

Face recognition, Bruce and Young say, involves a sequence of stages. In the first stage, we integrate the many views that we have seen of someone's face into a single image in memory, sort of like a prototypical image that best represents that person. From this facial image, we break it down into its component features and show how these features are arranged to form a specific face, much like a blueprint for constructing a house. Just as the plans for a home show the correct alignment of windows and doors, the blueprint of a face itemizes each of the necessary features, including the eyes, nose, and mouth, and shows how they fit together, forming one face that is distinguishable from all others.

Next we access knowledge in semantic memory to make attributions about the facial blueprint in terms of its general characteristics, such as its gender and age, and we search episodic memory to determine where and when we encountered this face before. The more times we have encountered someone's face, the stronger its blueprint and the more links that we have to it in our memory. This is why a friend's face seems to pop out in a crowd, and why, if a friend shaves off his beard, you are likely to notice a change, but mistakenly guess that he cut his hair or lost weight.[8] Losing a beard does not alter a facial blueprint because a beard is not a defining facial feature.

Face-Recognizing Machines

Computer scientists have applied similar ideas to face-recognition programs designed to search crowds for criminals or terrorists. Cameras guided by these programs, called *biometric face recognition systems* for the biological measurements they obtain, scan faces for features such as the distance between both eyes, the length of the ears, and even the texture of the skin. After a face is broken down into its essential elements, it is reconfigured into a unique mathematical description—called a *faceprint*—and compared to other face descriptions in the computer's memory, searching for an identifying match.

Steven Spielberg's science fiction film *Minority Report* features an omniscient version of machine recognition. Talking billboards in this film identify passersby from concealed cameras to present them with personal advertisements. One character, walking through a mall, has a billboard call out to him by name, saying, "Stressed out, John Anderton? Need a vacation? Come to Aruba!" Such ads may one day exist. Shopping malls in Tokyo already feature billboards that alter their displays to match the gender and approximate age of people walking by.[9] Face recognition technologies could soon render *Minority Report's* talking billboards eerily prophetic, raising issues over personal privacy.[10] For now, people are generally more versatile than machines at recognizing familiar faces, and some individuals are remarkably good, even with unfamiliar faces.

Exceptional Recognizers

Do you remember the people you saw yesterday at the supermarket or the person who sat beside you last week in the theater? Normally we find it easy to recognize family and friends, but we have difficulty recognizing briefly seen strangers. Studies of eyewitness accuracy show how fallible our memories can be, even if we are confident about our recollections.[11] But some individuals have an uncanny ability to recognize faces, even faces seen briefly at an earlier time. Although few in number, these individuals are so good at recognizing faces that they have been called *super-recognizers*.

When asked to identify childhood photos of famous adults or highly degraded images of briefly seen novel faces, the super-recognizers perform virtually perfectly, far above the level of age-matched adults. Their only drawback seems to be that they remember faces too well, requiring them to alter their social encounters with people who fail to recognize them. Said one super-recognizer, "I do have to pretend that I don't remember people … because it seems like I stalk them, or that they mean more to me than they do when I recall that we saw each other once walking on campus four years ago."[12] Those of us with more typical face-recognition ability would likely have the same reaction—Why would a total stranger recognize me, unless I was especially interesting to that person?

The super-recognizers show us that face recognition is not an all-or-nothing ability. It is an important evolutionary adaptation that enables us to distinguish friend from foe and lover from stranger. Most of us are good at it; some of us are exceptionally good, whereas others, as a result of brain damage or congenital defect, can be quite poor.

Have We Met Before?

One way to learn about person recognition is to observe what happens when this process breaks down. For some people, recognizing faces is a problem—a problem so severe that they are unable to recognize their own face.

The Stranger in the Mirror

Can you imagine not recognizing your own face? People who have difficulty recognizing faces suffer from *prosopagnosia*, a medical term loosely translated from its Greek roots as *face blindness*. It can be caused by a cerebral stroke, a traumatic brain injury, or a genetic defect present at birth that can affect members of the same family. The damaged brain areas most often involve the occipital lobes and the lower portion of the temporal lobes, including a structure buried deep inside called the *fusiform gyrus*.[13] What is peculiar about this disorder is that people with prosopagnosia are not blind. They can see. They can read and write and draw copies of pictures they are shown. They can look at a face, describe its features, usually know whether it is male or female, young or old, and correctly describe whether it is smiling or frowning. What they cannot do is identify a face.

To get a sense of this strange impairment, imagine looking at pebbles strewn on a beach. The pebbles are easily seen, and even though each one is different, it is hard to distinguish one from another. Face blindness is similar. Prosopagnosics say that all faces do not look alike; they just do not look like anyone in particular. As a result, unintended problems can arise with friends and coworkers. One woman with this disorder said that friends misinterpret her failure to recognize them as snobbishness when she passes them on the street without saying hello.[14]

Nevertheless, people with prosopagnosia can recognize others, just not by their face. They must rely on various distinctive cues such as a person's voice, hairstyle, clothing, or in some cases a distinctive walk. Sometimes, they recognize the people they know by context, as when two people agree to meet at a certain time and place. But danger lurks in these encounters. In arranging a meeting with her boyfriend, one prosopagnosic young woman had the following embarrassing experience: "I ran up to him and threw my arms around him and stretched up to kiss him; he drew back, pressing me away. It wasn't Dave. I had the wrong guy."[15]

It remains unclear whether prosopagnosia is a disorder specific to faces or a more general problem in object recognition. It may be that different recognition problems are linked to different forms of brain dysfunction. Psychologist Martha Farah, for one, has argued that recognizing faces and objects represent independent mental abilities because some individuals show general recognition problems for faces and objects, some have problems with objects but not faces, and still others only have problems identifying faces.[16] There is even one case of a man, CT, who was able to identify faces of people he knew from before a motorcycle accident, but unable to recognize new faces after his crash. When shown different photos of Michael Jackson, the pop star who dramatically altered his facial appearance over time, CT readily identified photos of Jackson that were taken before CT fractured his skull, but not any taken of him after the accident when Jackson's appearance was altered.[17]

In spite of their inability to recognize faces, prosopagnosics can still function effectively. In one instance, ten prosopagnosic members of an extended family had problems in recognizing each other and even their own faces in photos, but these people had above-average intelligence and achieved successful careers, including jobs as varied as programmer, physician, geologist, and engineer.[18] The neurologist Oliver Sacks is a notable example of a successful prosopagnosic, but even he has had mind-boggling experiences. Said he, "On several occasions I have apologized for almost bumping into a large bearded man, only to realize that the large bearded man was myself in a mirror."[19]

You Look Like Someone I Know

A different recognition problem occurs in people suffering from the *Capgras delusion*. This disorder is named after the French psychiatrist Jean Capgras, who, in 1923, described the curious case of a woman who claimed that various acquaintances had been replaced by *doppelgängers*—physical doubles. Delusions are false beliefs that are strongly held even in the face of contradictory evidence, and people experiencing the Capgras delusion are convinced that an identical-looking impostor has replaced a close friend or relative. This disorder has been observed in people diagnosed with schizophrenia, as well as people with brain lesions caused by a head injury, often in areas involved in face recognition. Playing on this idea of doppelgängers, films such as Richard Ayoade's *The Double* and Denis Villeneuve's *Enemy* have look-alikes take over a character's life, while Philip Kaufman's science fiction film, *Invasion of the Body Snatchers*, featuring

Donald Sutherland and Brooke Adams, has alien impostors replace humans after they go to sleep.

"In an hour, you'll be one of us."
Invasion of the Body Snatchers

Four friends living in San Francisco are deeply troubled. They have come to the outlandish conclusion that everyone else in the city is an impostor. They look like the people they always knew and they possess the same memories, but something about them is terribly wrong. Elizabeth Driscoll, for one, becomes increasingly distraught by an abrupt change in her boyfriend's behavior. Geoffrey used to be impulsive and passionate, but overnight he turned cold and distant. Questioning her own judgment, Elizabeth confides in her close friend, Matthew Bennell, saying,

> I know this is going to sound insane. Geoffrey is not Geoffrey. I mean, on the outside, Geoffrey's still Geoffrey, but on the inside, I can tell there's something different. Something is missing. What? Emotion, feelings, he's just not the same person.

Is Elizabeth suffering from paranoia? Has she fallen victim to the Capgras delusion? Initially scoffing at such talk, Matthew gradually changes his mind after hearing similar stories from others around the city. At the drycleaners, the shop owner tells Matthew that something is wrong with his wife, saying she is "not right." Later, an anguished woman runs from her husband at a party, crying out, "He's an imposter!" People all over San Francisco are having these strange thoughts. Their source, Matthew and Elizabeth discover, can be traced to aliens from outer space that arrived in the form of seedpods. Whenever a human goes to sleep, a pod replicates that person. But while the aliens are identical physically to the humans they snatch, they are devoid of human emotions, a dead giveaway about their alien identity.

Who is a person and who is a pod? *Invasion of the Body Snatchers* warns us not to take others at face value. The pod people have the right physical features and memories, but lacking the appropriate psychological characteristics, they are readily revealed as fakes. After initially trusting in appearance and memory, Matthew and Elizabeth soon realize that these are unreliable recognition cues, focusing instead on more intangible yet enduring psychological features. This film, a remake of Don Siegel's classic chiller of the same name, illustrates the importance of psychological characteristics for recognition, characteristics that we are normally only aware of when the people we know act "out of sorts."

Making Sense of a Delusion

An emotional twist links the aliens in *Invasion of the Body Snatchers* and people with the Capgras delusion. In the film, acquaintances were replaced by physical doubles that expressed no emotion; in the Capgras delusion, a person thinks others are imposters because the person feels no emotion toward them. Neuroscientist Vilayanur Ramachandran offers an explanation, based on an intriguing account of a young man who suffers from this delusion. After a traffic accident damaged the right side of his brain, DS could identify his father when speaking with him on the phone, but he steadfastly refused to believe it was his father when he saw him in person. A strange dialogue took place when Ramachandran asked DS why he thought his father was an imposter:

> DS: He looks exactly like my father but he really isn't. He's a nice guy, but he isn't my father.
>
> VR: But why was this man pretending to be your father?

DS: That is what is so surprising, why should anyone want to pretend to be my father?

This conversation resembles the anguished woman's comments in *Invasion of the Body Snatchers* who believed her husband was an imposter. How can DS accept his father's voice as real, but not his physical appearance? The answer, according to Ramachandran, may be found in selective damage in the brain's neural pathways that link areas that are important for our senses, emotions, and memories. For a Capgras patient, the link from the brain's voice recognition center to the brain's emotion center, the *amygdala*, works fine, but the connection from the face recognition center to the amygdala may be broken. Thus, the patient hears his father perfectly and has an emotional response to his voice, but although he sees his father's face and recognizes his appearance, there is no emotional reaction to the face.[20] In effect, the patient, staring in disbelief, says, "If you were my father, I'd feel something. But I don't, so you can't be him. You're an impostor!" In this instance, the imposter belief is delusional, but it affirms what we know about person identification. Someone's physical appearance is not the only element that we use in recognizing others, a point made by each of the films in this chapter, including Clint Eastwood's *Changeling*, featuring Angelina Jolie and John Malkovich, based on a true story involving a 1928 kidnapping in Los Angeles.

"I would know my own son."
Changeling

Returning home one afternoon after work, single mother Christine Collins discovers that her nine-year-old son, Walter, is missing. After frantically searching in vain, she calls the police who, several months later, bring home a boy that they claim is her son. But there is one small problem—Christine says that he is not her son. When Christine challenges the police by claiming they knowingly brought home an imposter, merely to protect their reputation, the authorities respond by questioning her sanity and placing her in a mental hospital. Obviously something must be wrong with Christine if everyone else believes that the boy is her lost son.

> Captain Jones: Mrs. Collins, listen to me. I understand. You're feeling a little uncertain right now and that's to be expected. Boys this age, who change so fast, we've compensated for that in our investigation, and there's no question that this is your son.
> Christine: That is not Walter.

Unlike Nicholas's mother who uncritically accepted Frédéric Bourdin as her lost son in *The Imposter,* Christine is convinced from the beginning that an impostor has been substituted for her child, even as others around her are perplexed and openly question her sanity. But Christine is not suffering from the Capgras delusion. In spotting obvious flaws in her "new" son, Christine shows that it is hard to fool Mom when the substitute son is shorter than her real son and does not know any of his former schoolmates. In lacking the requisite physical features, psychological characteristics, and shared memories, the imposter turned out to be a boy who had run away from home and was looking for somewhere to live. There was no way that Christine could accept him as her own. She saw with her eyes, not her heart.

Fade-Out

Intuitively, we place great faith in a person's physical features, but as the films in this chapter show, we should not be so quick to accept others at face value. The less obvious elements of personality and shared memories offer additional cues for recognizing the people we know. A change in any of these elements signifies a fundamental change in another person. This is how the characters in this chapter's films eventually make the right recognition decision and why it is so difficult to live someone else's life. It is simply too hard for an imposter to get all three of these essential elements right.

5 Autobiographical Memories and Life Stories

Setting the Scene

When recalling personally experienced events, we call on our autobiographical memories—the memories that define us and make us unique. These personal memories are molded into a coherent life story that shapes our individual identity. But in forming a life story, we function as memoir writers, interpreting and reinterpreting our past, to create a reconstructed narrative. The characters in this chapter's films illustrate this narrative aspect of memory by telling their life stories through flashbacks.

Featured Films

The Kite Runner, directed by Marc Foster, with Khalid Abdalla, Zekira Ebrahimi, and Ahmad Khan Mahmoodzada (DreamWorks Pictures, 2007, English and Dari with English subtitles, Rated PG-13).

Slumdog Millionaire, directed by Danny Boyle, with Dev Patel, Freida Pinto, and Madhur Mittah (20th Century Fox, 2007, Rated R).

Cinema Paradiso: The New Version, directed by Giuseppe Tornatore, with Salvatore Cascio, Philippe Noiret, and Jacques Perrin (Miramax Films, 2002, Italian with English subtitles, Rated R).

Titanic, directed by James Cameron, with Kate Winslet and Leonardo DiCaprio (Paramount Pictures and 20th Century Fox, 1997, Rated PG-13).

Unchained Memories: Readings from the Slave Narratives, documentary directed by Ed Bell and Thomas Lennon (HBO Video, 2003, Unrated).

The Joy Luck Club, directed by Wayne Wang, with Ming-Na Wen, Tsai Chin, Rosalind Chao, and Tamlyn Tomita (Buena Vista Pictures, 1993, English and Mandarin with English subtitles, Rated R).

Narrative Flashbacks

Do you remember when you first fell in love? Recalling your first romance requires *autobiographical memory*—the ability to remember personally meaningful events that are often self-referential, interpersonal, and emotional. In recalling that past romance, you likely remembered multiple episodic events including your first meeting and the places you went together, along with various bits of semantic knowledge such as your partner's name and eye color. All it takes is the right cue to trigger a recall, and you become a memoirist telling a story.

Writer Marcel Proust's memoir, *À la recherche du temps perdu* (*In Search of Lost Time*), provides a vivid example of this triggered recall. His long-dormant childhood memories came flooding back after he tasted a morsel of French pastry, a *petite madeleine* dipped in a cup of tea:

> The taste was that of the little piece of madeleine which on Sunday mornings at Combray ... my aunt Léonie used to give me. ... And as

> soon as I had recognized the taste ... the whole of Combray ... sprang into being, town and gardens alike, from my cup of tea.[1]

Proust's voluminous recollections provide irresistible descriptions of autobiographical recall—the same type of remembering shown in film by a *flashback* when a scene from the present fades into a scene from the past. Signaled by a change in music, a memory object such as a photograph, or dialogue as in Marc Forster's film *The Kite Runner,* filmmakers use flashbacks to tell viewers that the story is shifting backward in time to reveal how the present was shaped by the past.

"There is a way to be good again."
The Kite Runner

The opening of *The Kite Runner,* featuring Khalid Abdalla, Zekiria Ebrahimi, and Ahmad Khan Mahmoodzada, shows children flying kites over a San Francisco park, while a young writer named Amir Qadiri watches and remembers. Returning home with his wife, Amir receives an unexpected call from Rahim Khan, an old friend of his deceased father who urges him to return to Afghanistan, adding, "You should come home. There is a way to be good again." With these ominous words, the flashback begins and we embark on a memory journey, back in time to Amir's youth in Kabul when he flew kites with his best friend, Hassan.

Based on Khaled Hosseini's novel, *The Kite Runner* tells a story about guilt and redemption—about correcting a wrong that once done had unforgettable consequences. One fateful day, with the sky above Kabul filled with kites, boys all over the city competed in a kite-fighting contest, slicing each other's kite to the ground. Eventually, one kite remained—the one flown by Amir and Hassan. The son of Amir's family servant, Hassan excelled at kite running, knowing exactly where each kite would land. Running for the last kite, Hassan is followed by the neighborhood bully, Assef, who orders him to give up the kite. Refusing, Hassan is beaten and raped, while a frightened Amir watches and hides. Riddled with guilt, Amir ends their friendship, but he never forgets what happened to his friend, even after he immigrates with his father to America.

Flashing forward two decades later, an adult Amir visits the dying Rahim Khan who reveals that after the Taliban seized control of Kabul, they executed Hassan and his wife for protecting Amir's home and took their son, Sohrab, to pleasure their brutal leader, Assef. Saddened by this news, Amir is then stunned to learn that Hassan was his half-brother and Sohrab is his nephew. Urging Amir to find the boy, Rahim hands him a letter written by Hassan, describing his dreams and everlasting friendship:

> I dream that my son will grow up to be a good person. ... Flowers will bloom in the streets of Kabul ... and kites will fly in the skies. And I dream that someday you will return to Kabul. ... If you do, you'll find an old faithful friend waiting for you.

Reading these heartfelt words, Amir knows that fate has offered him a second chance. Now he has the opportunity to be good again.

Moving backward and forward in time, *The Kite Runner* demonstrates how a character's autobiographical memories create a coherent life story, virtually the same type of process that occurs when we tell others the stories of our lives.

Tell Me a Story about You

"If you want to know me," says psychologist Dan McAdams, "then you must know my story, for my story defines who I am."[2]

The stories that we tell of our lives often follow a narrative script involving people, places, and events in which we play the leading role.

Life Narratives

Beginning during childhood, our life stories take shape as we organize our experiences—whether the first day of school or a trip to the zoo—into meaningful recollections that we can tell other people. Using personally experienced events, we systematically mold and shape these memories into a life narrative. When recalling a trip to the zoo, for instance, we describe that event by imposing a storylike structure on that recollection because stories are easy to tell. But in telling this first-person drama, we can transform it by omitting unflattering details, embellishing others, and filling in missing pieces to give the event coherence and meaning. The result is a reconstructed memory narrative, part fact and part fiction, that becomes part of our past over time. But, as McAdams notes, we are more than just tellers of tales:

> We each seek to provide our scattered and often confusing experiences with a sense of coherence by arranging the episodes of our lives into stories. ... We are not telling ourselves lies. Rather ... we compose a heroic narrative of the self that illustrates essential truths about ourselves.[3]

Forming a life narrative does not mean that our personal memories are self-serving fictions, designed to make us look good as the protagonist of our story. Life narratives evolve and change with new experiences, but they remain more fact than fiction even though we reconstruct our memory of the past. Filmmaker Sarah Polley found this to be true in her documentary, *Stories We Tell*, as she attempted to learn about her deceased mother through the disparate recollections of family and friends. Our memory is imperfect because we rarely keep precise records of what happened, but we remember the gist of our experiences well. People who recorded daily events in a diary, for example, were able to recognize their own entries several months later, as well as altered versions of their entries that preserved the gist of those events. However, these same people were much less inclined to accept someone else's diary entries as their own.[4] Our autobiographical memories usually provide a more or less accurate record of our life experiences, a generally faithful guide to our past that is necessary for our survival as a species.

Remembering Lifetime Experiences

When I was a boy of ten, I loved playing baseball. I grew up in southern Connecticut, an hour from New York, when the city was home to the Yankees, Dodgers, and Giants. Most of my friends followed one of these teams in the 1950s, but I adopted a team from the Midwest. My hero was a young slugger named Eddie Mathews who played third base for the Milwaukee Braves. One day, when the Braves came to New York, my father took me to the Polo Grounds to watch them play the Giants. We sat next to the third base railing, and while I have no memory of the game, I recall vividly what happened after the final out. Back then, fans could leave the ballpark by the field exits and the players would walk out to the clubhouse in center field. As we were leaving, I spied my boyhood idol walking up ahead. I raced up to him, wanting to say a million things, but could only manage to stick out my hand and timidly say, "Nice game, Eddie." He gave my outstretched hand a brief squeeze and replied, "Thanks, kid." As he walked away, I found my father in the crowd, smiling. Nice game, Dad.

This personal recollection reveals how I structured this autobiographical memory. In recalling a long-ago ballgame, I described three kinds of autobiographical knowledge. First, I described a *lifetime period* by telling about my youth. Next, I described a *general event* by describing my love of baseball and the teams that my friends and I followed. Finally, I shared an *event-specific memory* by describing a particular event at a specific time and place—the day that I spoke to my boyhood idol at the Polo Grounds. Over a lifetime we forget many event memories, such as the train ride I took that day to New York, while others endure when they are unique and colored by emotion—this awe-struck kid's special moment. We use all three types of knowledge, says psychologist Martin Conway, in telling our personal stories.[5]

Lifetime periods are measured in years. They give structure to our autobiographical memories by organizing episodes by a common theme, such as those times playing with childhood friends or courses taken in college. General events, measured in days, weeks, or months, are more specific than lifetime periods. They represent a sequence of events that form a larger episode, such as working at a summer job, or repeated events, such as attending high school algebra class. Finally, event-specific memories are episodic memories of single events, such as opening a college acceptance letter or observing the birth of a child. These memories are measured in seconds, minutes, or hours.[6] The three levels of autobiographical knowledge not only guide our narrative recall, they also provide the narrative structure for Danny Boyle's *Slumdog Millionaire*, in which a young man's personal memory leads to unimagined wealth.

"So are you ready for the final question for 20 million rupees?"
Slumdog Millionaire

This fairy tale of a film, featuring Dev Patel, Freida Pinto, and Madhur Mittah, tells the story of eighteen-year-old Jamal Malik, an impoverished tea-server on the brink of winning India's version of *Who Wants to Be a Millionaire*. All he needs to do is risk everything he has already won and answer one more question correctly. But something is strangely out of whack here. How did an uneducated "slumdog" orphan, raised in dire poverty on the streets of Mumbai, acquire the broad knowledge needed to reach the final round of India's favorite game show? That is what the producers of the show wonder, as Jamal becomes a television sensation with 60 million viewers tuning in to see if this pauper can turn into a prince. Even film viewers are asked to ponder:

> Jamal Malik is one question away from winning 20 million rupees. How did he do it?
>
> A. He cheated
> B. He's lucky
> C. He's a genius
> D. It is written

To find out, Jamal is arrested on suspicion of cheating and questioned and tortured by the police. It is during these interrogations that Jamal relays his life story, using a series of flashbacks to move from the present to the past, showing how a rag-tag team of three orphans—Jamal, his brother Salim, and a girl named Latika—learned to survive by their wits in Mumbai with little formal education. Through these flashbacks when Jamal describes his experiences growing up, he reveals how he learned the answers to the game show questions.

Seated on the set of the show, with the audience and game show host looking on, Jamal is read the first question: "Who was the star of the 1973 hit film *Zanjeer*?" As Jamal concentrates, his mind flashes back to when he was a young boy in the slums of Mumbai—a *lifetime period*—renting toilet stalls with his brother—a *general event*—while waiting for the helicopter arrival of Amitabh Bachchan, the star of the action film *Zanjeer*. Locked in a stall by his brother, Jamal escapes by dropping into a pit of human waste, running out covered in feces, while holding a small photo of Amitabh that he kept safe in his pocket. Disgusted fans curse as Jamal pushes his way through the crowd, finally reaching his hero and holding up the photo for an autograph—an *event-specific memory*.

> Jamal: Amitabh-ji! Amitabh-ji! Please. Amitabh-ji!
> (Amitabh takes it, scribbles his name, and is hustled away by his bodyguards.)
> Jamal (kissing the photo): Yea! A thousand thanks, Amitabh-ji.

Indelible memories like this one provide Jamal with the answers he needs.

Slumdog Millionaire speaks to the various functions of autobiographical memory, including how we use past experiences to solve current problems and how these experiences help shape our identity.[7] Jamal's experiences growing up in Mumbai, lying dormant in his autobiographical memory until needed, bring him to the game show's ultimate question. But Jamal's goal was never to achieve great wealth or fame. Like all good fairy-tale heroes, he wisely followed his destiny by seeking something that wealth or fame could not buy. How did he do it? One answer alone is correct.

Emotional Movies and Empathy

The Kite Runner and *Slumdog Millionaire* touch us emotionally. We are saddened watching Amir reading a letter from his deceased friend Hassan, but gladdened seeing Jamal answering questions that should stump him. We know these are fictional characters, but it does not matter—we react to what we see on the screen.

Film viewers once feared taking showers after watching a brutal murder scene in Alfred Hitchcock's *Psycho*, and they avoided dips in the ocean after seeing a shark terrorized beachgoers in Steven Spielberg's *Jaws*. Watching George Romero's scary *Night of the Living Dead* has even led viewers to yell out in a tension-filled theater, "Don't go out there!" It is only a movie, yet we are moved by the experience—almost like real life.

How Movies Move Us

How do movies gain such a grip on our feelings that we can burst out laughing in one scene and fail to hold back tears in another? Where does the empathy that we feel toward Amir and Jamal come from? The answer may lie in our brain's *mirror neurons*, part of the cortical networks that are active whenever we act or experience an emotion as well as when we observe someone else acting or responding emotionally.[8] Stories, as noted previously, function as our mind's flight simulator, allowing us to glimpse life through a character's eyes, enabling us to imagine how we might navigate similar challenges in the future.[9] An important part of this navigation process is sharing an emotional experience, mirroring the experience of others to feel what a character feels.

Mirror neurons may play a role in this process. Brain scans reveal that similar brain areas are activated when people experience an emotion or perform an action as when they see the same emotion or action in film.[10] Simulating events in our brain helps us understand a story, and mirror neurons add emotional empathy, says neuroscientist Marco Iacoboni, by recreating in our brain the emotions shown on the screen:

> We have empathy for the fictional characters—we know what they're feeling—because we literally experience the same feelings ourselves. And when we watch the movie stars kiss on the screen? Some of the

cells firing in our brain are the same ones that fire when we kiss our lovers.[11]

Cortical networks involving mirror neurons can help explain the empathy we feel toward Amir and Jamal. Their life stories are emotional, just as our own autobiographical narratives often are when we recall our lifetime experiences.

Emotional Links to the Past

When people are given a list of common words and asked to recall the first memory that comes to mind for each word, they are apt to report more recent than distant memories. Try it for the cue word *breakfast*. You likely recalled what you had for breakfast this morning. People also report a greater sense of recapturing the past with an emotional cue word such as *happiness* than a neutral word such as *river*. But regardless of cue, our memories seem to fade with the passing of time, as do our emotional reactions to these remembrances.[12] Oftentimes, we demonstrate a Pollyanna effect by recalling pleasant events better than unpleasant events.[13] This positive bias occurs partly because negative memories seem to fade faster than positive memories, a phenomenon called the *fading affect bias*.[14] The exception, covered in the next chapter, is the clear recall we typically show for traumatic events such as combat experiences or the loss of a loved one.

Many films tell autobiographical stories that are colored by deeply felt emotional memories. Films such as Sydney Pollack's *Out of Africa*, Anthony Minghella's *The English Patient*, and Steven Spielberg's *Saving Private Ryan* use flashbacks to tell their tales about a memorable time in a character's life. Their characters are all emotionally tied to their past, but none more so than Salvatore DiVita in writer and director Giuseppe Tornatore's *Cinema Paradiso: The New Version*, with Salvatore Cascio, Philippe Noiret, and Jacques Perrin.

"He left something for you."
Cinema Paradiso

Cinema Paradiso is an affectionate tribute to autobiographical memory and film. The title refers to the name of a small theater in a Sicilian village where, before the advent of television, young and old alike learned about life from the movies. Salvatore DiVita spent much of his youth in this theater, but he left as a young man, going on to make his own movies and becoming a famous director in Rome. After receiving a late-night call from his aged mother, informing him that his former mentor, Alfredo, has died, Salvatore reminisces, flashing back thirty years in his mind. What follows is a memory play in three acts.

Act 1 shows Salvatore as a young boy called "Toto," developing his love of film with Alfredo, the theater's projectionist who dispenses advice with movie quotes. It does not take long for the childless Alfredo and the fatherless Toto to bond.

In act 2, a decade later, a teenage Salvatore is the projectionist after a fire from an overheated projector has blinded Alfredo. But movies are no longer Salvatore's sole interest. He has fallen in love with Elena, a girl whose parents have higher ambitions for her and ultimately take her away—but not before she promises to meet Salvatore in the theater, a rendezvous that never takes place. Desolate, Salvatore turns to Alfredo, who urges him to move on, saying, "Life's not like you saw it in the movies. Life is harder." With no word from Elena, Salvatore heeds his mentor's advice, until the funeral brings him home for act 3 of this memory drama.

Flashing forward, a middle-aged Salvatore returns, nostalgically viewing the people and town. So much is different, including the once bright Cinema Paradiso that is now deserted and neglected, waiting for demolition. But not everything has changed. By chance, Salvatore encounters Elena's teenage daughter, a young woman who is the image of the Elena of his youth. Following her home, he locates the now older and married Elena, the love he never forgot. Hesitantly phoning her one evening, Salvatore requests a meeting, but Elena declines, saying, "It's been so long ... What good would it do?" Yet, she surprises him that night, showing up on the beach road where they used to park, calling Salvatore into her car, and setting off an emotionally charged reunion. Still hurting after so many years, Salvatore cannot resist asking Elena why she never kept their long ago date.

> Salvatore: The last time we saw each other, we made a date to meet at the Cinema Paradiso ... And you didn't come, you disappeared without leaving a trace, nothing!
> Elena: I kept that date. But I was late ... I had a fight with my family.

Sadly, Salvatore learns that Elena left him a note that evening, a note still hanging on the projection room wall. Tearful over what might have been, they consummate their passion in the cramped car, but this will be their only reunion. When Salvatore eagerly phones the next day, Elena gently interrupts, "No Salvatore, there is no future. There's only the past. Even meeting last night was nothing but a dream, a beautiful dream."

With Elena's tender rebuff, Alfredo's funeral, and the demolition of the Cinema Paradiso, Salvatore's emotional ties to his past are coming undone. The people who were once the center of his life are now only ghosts, poignant reminders of what was. It is time to let the past go. Almost. Following the funeral, Alfredo's wife gives Salvatore a gift that Alfredo has kept for him, a gift sure to get your mirror neurons firing. We experience it with a teary-eyed Salvatore as he watches one of the most sensual collections of romantic moments ever collected on film.

Distant Memories

When does our memory of the past begin? Try recalling your earliest specific memory and note your age at the time. When college-aged and elderly people respond to this question, most report nothing before the age of two, followed by fragmented recollections during their third and fourth years.[15] Tests involving verifiable events from infancy and early childhood, such as the birth of a younger sibling or an overnight hospital stay, confirm the difficulty that adults experience in recalling these distant events.[16]

A Forgettable Time

The universal inability to recall early life experiences is called *childhood amnesia*, and it is likely due to several factors. First, after birth there is continued brain maturation in the frontal lobes and hippocampus, neural structures that are important for long-term retention. Second, cognitive development occurs as children acquire a rudimentary understanding of thought over time. One day, for example, while I was riding in a car with my nearly four-year-old granddaughter, she turned to me and asked, "What are you thinking?" I was amazed by the complexity of the question. It implied that she thought, that she understood that other people thought, and that we could share our thoughts. She was developing a theory of mind. She was also developing autobiographical memory, reminding me that I had missed her last birthday party.

Language skills also are acquired in childhood, and autobiographical memory development is closely linked to language development. When people are given cue words to recall specific memories, the earliest memory associated with each cue word trails the acquisition of that word during childhood by several months—in effect, first words precede first memories.[17] Finally, culture plays a role. Young adult Euro-Americans produced earlier first memories than similarly aged Taiwanese by a year or more, owing to cultural differences in socialization, including the ways in which parents discuss the past with their children.[18] In later childhood, a shift takes place, with autobiographical recollections taking on a more adultlike narrative structure, and by adolescence, individuals begin recalling events as episodes in their life story.[19]

A Most Memorable Time

Contrasting with our inability to remember experiences from infancy and early childhood is our heightened ability for recalling experiences from another time. To see which time period is especially memorable, complete each of these statements:

For the most important positive event of my life, I was ____ years old.

For the time I was most in love, I was ____ years old.

When do these experiences most often occur? Adults from their twenties through their seventies typically report the same thing—they most often occur around age twenty. Positive experiences that occur during late adolescence and early adulthood are remembered well into old age, in cultures all over the world.[20] Whether recalling personal events such as a romantic relationship or general knowledge such as a major news story, people show heightened memorability for experiences surrounding age twenty, a phenomenon that psychologist David Rubin calls the *reminiscence bump*.[21] This bump is not the result of greater reminiscence of these memories; it is due to major lifetime transitions that involve changes in our self-image.[22] In many cultures around the world, a major transition followed by a period of relative stability occurs during late adolescence and early adulthood when culturally shared positive experiences, such as completing school and beginning a career, typically take place. Memories for negative experiences, whether becoming ill or losing a loved one, are not linked to any lifetime period and show no age-related memory bump.

Looking back to this section's opening statements, you know that the time you were most in love, most likely during late adolescence or early adulthood, has not been forgotten. These intense romances stay with us, producing a reminiscence bump that lasts a lifetime. Writer and director James Cameron uses this aspect of romantic reminiscence as the narrative foundation for his film, *Titanic,* featuring Leonardo DiCaprio and Kate Winslet. Rose Dawson Calvert is a month shy of being 101 years old, yet she has never forgotten her brief, but life-altering romance with twenty-year-old Jack Dawson that happened when she was seventeen.

"It's been eighty-four years, and I can still
smell the fresh paint."
Titanic

The historical facts surrounding the sinking of the *RMS Titanic* during its maiden voyage are familiar to most. Less than three hours after striking an iceberg, the ship that was considered unsinkable sank to the bottom of the North Atlantic on April 15, 1912, and 1,502 lives were lost at sea. But knowing of this epic disaster and feeling it in human terms are different; *Titanic* helps us get it by using the maritime tragedy as the backdrop for the fervent romance of two fictional characters.

Titanic connects the present to the past by taking viewers down to the ocean bottom, where treasure hunters—searching the *Titanic*'s cold, silent remains—retrieve a safe believed to contain a blue diamond necklace, called the Heart of the Ocean. Topside, they find that the safe's only artifact is a sketch of a nude young woman, wearing the priceless necklace. Seeing the sketch on the local news, an amazed Rose Calvert phones Brock Lovett, the lead treasure hunter, who asks if she knows the identity of the woman in the drawing. "Oh yes," says Rose, "the woman in the picture is me." Skeptical, Brock flies Rose out to his ship, seats her before an array of monitors revealing the *Titanic* resting 12,000 feet below, and asks if she is ready to go back. Staring at the haunting undersea images, Rose's memories return. We are all going back to the *Titanic*.

Rose reveals that she boarded the ship as Rose DeWitt Bukater, traveling first class and engaged to a wealthy snob for the sake of her widowed, penniless mother. Unhappy with her life and its stifling, aristocratic lifestyle, Rose stands irresolute on the rail of the ship's stern, ready to jump, until a young artist pulls her to safety. Jack Dawson is homeless and poor, having won his ticket in a card game, but he is the free spirit that Rose longs to be. When he invites her to a third-class party, full of spirited dancing and drinking below deck, Rose accepts, much to her mother's dismay. Rose is now taken with Jack. Impressed with his drawings, she asks him to sketch her, wearing only the extravagant necklace. As Rose lies naked on a chaise, a nervous Jack sketches, peeking over the top of his pad. When he finishes, Rose glances approvingly over his shoulder, before stowing the drawing in a safe, where it will remain for eighty-four years.

Thoroughly outraged, Rose's fiancé sends his bodyguard out to retrieve her, but the lovers elude him by hiding in a car in the ship's stowage. Posing as a chauffeur, Jack asks, "Where to, Miss?" Rose, yanking him into the back seat, replies, "To the stars!" Their passion is short-lived, once the inevitable iceberg arrives.

Choosing Jack over the safety of a lifeboat, Rose grasps his hand as the ship sinks into the frigid sea. Amid the cries of people around them, Rose lies shivering on a wooden plank, while Jack, clinging to its side, makes her promise that she will survive. She does, kissing his hand before he slips underwater, just before a lifeboat arrives.

Flashing forward to the present, the elderly Rose sees that Brock and his rapt crew now get the human tragedy that was the *Titanic*. Concluding, she says:

> But now you know there was a man named Jack Dawson and that he saved me in every way that a person can be saved. I don't even have a picture of him. He exists now only in my memory.

Rose was seventeen when she met the love of her life, a period of late adolescence and early adulthood when we are most in love and our memories are most enduring, even over eighty-four years.

Unforgettable Moments

For the actual survivors of the *Titanic*'s sinking, it was an unforgettable night. Said one seventy-eight-year-old survivor who was eight years old at the time, "It seems to me that I heard an explosion ... steam, smoke, fire, and flashes, and then it was gone. Then there were the cries."[23] Indelible memories such as this one are called *flashbulb memories*, the seemingly clear, confident, and persistent recollections that people have of momentous events. The sinking of the *Titanic* was one such event; the assassination of President Abraham Lincoln was another. Thirty-three years after Lincoln was assassinated in 1865, people were asked if they recalled where they were when they learned that Lincoln was shot. A large majority of people not only recalled where they were; they were able to provide detailed information regarding how they heard the news, what they were doing, and how they felt at the time.[24] Similar detailed personal recollections have been reported following other momentous events including the assassination of President John Kennedy, the Space Shuttle *Challenger* disaster, and, more recently, the attack on the World Trade Center in New York.[25]

Are flashbulb memories a uniquely accurate type of autobiographical memory that is immune to forgetting over time? It turns out that these highly confident recollections often contain inaccuracies. When detailed records are made of people's memory immediately after a momentous event, such as the morning of September 11, and later compared with their recollections of the same event months or years later, there are oftentimes numerous changes between the first and most recent remembrances. People remember the terrorist attack, but they will confidently misremember details such as where they were at the time or how they first heard the news. Even President George W. Bush showed inconsistencies in recalling how he learned of the attack over time.[26]

Yet not all flashbulb memories show inaccuracies. People who personally experience a momentous event, whether surviving an earthquake or a terrorist bombing, remember the event accurately over time, whereas others who merely experience the event in the news do not.[27] The same may be true of uniquely personal events such as the birth of a first child or the loss of a loved one, something you deeply experienced just once. Our flashbulb memories for distinctive events are much discussed and rehearsed, and, over time, these autobiographical recollections take on a story narrative, similar to a news story that describes a momentous event by reporting who, what, where, and when. Moreover, the older the event remembered, the more we visualize ourselves as part of the memory, rather than seeing the event from our original perspective.[28] But that is not how we originally perceived the event; it is how we reconstructed the event from our memory. Autobiographical memories are personal interpretations of events, not snapshots or recordings of events.

Using the term "flashbulb" to describe these autobiographical memories is misleading, according to psychologist Ulric Neisser, because "such memories are not so much momentary snapshots as enduring benchmarks. They are the places where we line up our own lives with the course of history itself and say 'I was there.'"[29] The year 1865 was especially notable for these historical benchmarks. Not only did people learn of Lincoln's shooting that year; many had an unforgettable moment created by something he achieved while alive.

"I could tell you about it all day, but even then
you couldn't guess the awfulness of it."
Unchained Memories

Ed Bell and Thomas Lennon's documentary, *Unchained Memories: Readings from the Slave Narratives*, provides an authentic historical benchmark in film. It describes America through 1865, when slavery still existed. Photographs from that time remain and, thanks to the Federal Writer's Project, so do the memories of former slaves. Following the Emancipation Proclamation in 1862 and the end of the Civil War in 1865, over four million slaves gained their freedom. Approximately seventy years later, during the 1930s, the memories from roughly 2,000 of these elderly former slaves were recorded, spoken as oral histories. Narrated by Whoopi Goldberg and with readings by African-American actors, this documentary uses their words to tell their stories, supplemented with photos from those times. These people are long dead, but their collective memories endure to tell us what happened—how they suffered and how they survived.

One day was particularly memorable for Katie Rowe. Working in the fields, she heard the blast of a horn and returned to her quarters with the other slaves to find an elderly white man waiting for them. He asked them if they knew what day it was, and when they replied that they did not know, Rowe recalls what the man said next:

> "Well, this is the fourth day of June and this is 1865, and I want you to all remember the date, cause you always going to remember the day. Today, you is free."

And then, she remembered, he got on his horse and rode off.

Whether that day served as a flashbulb memory or historical benchmark for Katie Rowe, it was an unforgettable moment. For her and the other former slaves, their autobiographical recollections are oral histories that are subject to all of the failings of memory—they can fade or be distorted over time, and they are influenced by what each person wants to report in an interview. Yet, collectively, the oral histories of former slaves, as well as those of Holocaust survivors—reenacted in films such Steven Spielberg's *Schindler's List* and Janet Tobias's *No Place on Earth*—provide consistent first-person descriptions of unforgettable historical events.

Remember Me

Oral histories are one way in which people pass on their memories to succeeding generations. Memoirs are another. In his book

The Last Lecture, Randy Pausch wrote a memoir for his children that grew out of a talk he prepared. Addressing a standing-room only crowd on September 18, 2007, this forty-seven-year-old computer science professor had pancreatic cancer and less than a year to live. Briefly acknowledging his illness, Pausch focused on his life narrative by describing how he had achieved each of his childhood dreams and the insights he picked up along the way.

What was most touching was his desire to leave lasting memories for his three children, ages six, three, and one, at that time. Pausch realized that their memories of him would be fuzzy, noting, "That's why I'm trying to do things with them that they'll find unforgettable." He swam with dolphins with his oldest son, Dylan, saying, "A kid swims with dolphins, he doesn't easily forget it." Middle child Logan got a trip to Disney World to meet Mickey Mouse. And for his infant daughter, Chloe, knowing she was too young to remember him, he left her this wish—"I want her to grow up knowing that I was the first man ever to fall in love with her."[30]

Ugandan mothers stricken with AIDS have used their memory in the same way as Randy Pausch. Christa Graf's German documentary, *Memory Books*, focuses on four Ugandan mothers who compiled their family's history in words and pictures. Their memory books are a way of being remembered while preserving values and traditions in their children after they are gone. Graf's documentary is difficult to obtain, but its focus on one generation's passing along its memories and hopes to the next is captured well in Wayne Wang's film, *The Joy Luck Club*, featuring an ensemble cast of Asian-American actors including Ming-Na Wen, Tsai Chin, Rosalind Chao, and Tamlyn Tomita.

> "This feather may look worthless, but it comes from afar and carries with it all my good intentions."
>
> *The Joy Luck Club*

Based on Amy Tan's novel and set in present-day San Francisco, *The Joy Luck Club* tells of four middle-aged women who were born in China before the 1949 Cultural Revolution, immigrated to America, and became friends. At their weekly mahjong meetings, a frequent topic of conversation is their daughters. When one of the women dies, her young adult daughter takes her place at the table. It is here that June hears stories from each of her "aunties" about life years earlier in China. It is also where she learns that her late mother had dreams and plans of her own, and that she has two adult half-sisters waiting to meet her in China.

The story begins at a large family gathering in June's home before she travels to China to meet her new sisters. Through a series of flashbacks, we learn what life was like for each of the mothers and what life is like for each of the daughters. This affecting film is about the emotional power of generational memory. It is about women who were raised in China and their thoroughly American daughters who know nothing about the past joys and heartbreaks of their mothers. This generational divide leads inevitably to mother-daughter conflicts, such as one between June and her late mother, Suyuan, that get resolved with shared understanding.

> June: I'm just sorry ... I've always been so disappointing ... I'll never be more than what I am. And you never see that, what I really am.
>
> Suyuan: You, your thinking different. Because you have best quality heart ... No one can teach. Must be born this way. I see you.

The Joy Luck Club is more than just a film about resolving strained relationships. It is about who we are now and how our identity is influenced by those who came before us—how one generation strives to pass on its knowledge to the next generation and how their life stories help define us. Whether it is a father dying of cancer, a mother dying of AIDS, or a mother born in a different culture, the ability to pass along memories and wisdom to our children, represented by the passing of a swan's feather from a mother to her daughter in *The Joy Luck Club,* may be autobiographical memory's most treasured function.

Fade-Out

Our autobiographical memories combine aspects of episodic and semantic memory into personal interpretations of events, not snapshots or recordings of events. We mold these reconstructed memories into a coherent narrative that we recall using lifetime periods, general events, and event-specific memories. Like film characters that tell their stories in flashback, our life stories define us, and our autobiographical recollections are generally faithful guides to our past.

6 When Troubling Memories Persist

Setting the Scene

Emotion enhances memory for both positive and negative experiences. Enduring positive memories provide pleasure, whereas persistent negative memories produce pain. In dealing with troubling memories, several characters in this chapter's films experience the debilitating symptoms of post-traumatic stress disorder, making it difficult to cope, while others overcome their toxic memories, achieving personal growth and change.

Featured Films

Remembrance, directed by Anna Justice, with Dagmar Manzel, Alice Dwyer, and Mateusz Damięcki (Corinth Films, 2011, English, German, and Polish with English subtitles, Unrated).

Born on the Fourth of July, directed by Oliver Stone, with Tom Cruise, Frank Whaley, Willem Dafoe, and Kyra Sedgwick (Universal Pictures, 1989, Rated R).

Rachel Getting Married, directed by Jonathan Demme, with Anne Hathaway, Rosemarie DeWitt, Bill Irwin, and Debra Winger (Sony Pictures Classics, 2008, Rated R).

Mystic River, directed by Clint Eastwood, with Sean Penn, Tim Robbins, and Kevin Bacon (Warner Bros. Pictures, 2003, Rated R).

Capturing the Friedmans, documentary directed by Andrew Jarecki, with Arnold Friedman and family (Magnolia Pictures, 2003, Not Rated).

Ordinary People, directed by Robert Redford, with Timothy Hutton, Donald Sutherland, Mary Tyler Moore, and Judd Hirsch (Paramount Pictures, 1980, Rated R).

Haunted by the Past

Imperfect though it may be, our autobiographical memory is generally faithful in remembering the emotional experiences of a lifetime, especially moments that touch us deeply. I can still clearly recall the physician calling out to me in a hospital delivery room—"John, come meet your new son!"—just as I also remember speaking at my mother's funeral. Everyone has memories colored by positive or negative affect, and remembering them is only troublesome if they interfere with everyday life. Whether we are consumed by a nostalgic longing for happier times or stressed by a disturbing event from long ago, the burden of carrying these memories can be taxing.

Films involving Holocaust survivors, such as Alan Pakula's *Sophie's Choice* and Stephen Daldry's *The Reader*, show how long-held troubling memories could dominate a character's life. Lacking resolution, these emotional memories can last for decades, as shown in Anna Justice's film, *Remembrance*, featuring Dagmar Manzel, Alice Dwyer, and Mateusz Damięcki. Inspired by true events, this film tells the story of a woman unable to forget her wartime past and the man she had intended to marry.

"I am haunted by memories that refuse to be forgotten."
Remembrance

Living a quiet life in 1976 with her husband and daughter in New York, Hannah Levine stops by the dry cleaners one day and her world is turned upside down. Catching bits of an interview airing on the shop television, she sees a middle-aged man describe how he met the love of his life—a young woman in a German concentration camp—before escaping with her in 1944.

> Interviewer: How was it even possible to fall in love in a concentration camp? Were you even allowed to speak with the girl?
> Man: No.
> Interviewer: So how did you communicate?
> Man: We spoke with our eyes first, later with food. She understood how much I loved her long before I spoke.

Now in her fifties, Hannah is stunned as the memories return yet again. Could the man she saw be Tomasz, the man she vowed to marry, the man who saved her life? Rummaging through an old folder, stuffed with unsuccessful attempts to find him, Hannah locates his ragged photo. Staring at the faded image, she reflects, "I thought I was finished with the past. Done. But you're never done." Three decades earlier, she was Hannah Silberstein, a German Jew, and the young man in the photo was Tomasz Limanowski, a Polish freedom fighter. They fell in love in the darkest of places and times.

Hannah scrubbed floors in the camp bakery, scavenging crumbs to survive, while Tomasz held a desk job, sharing food with Hannah in their stolen moments together. They escaped when Tomasz swiped a German officer's uniform and ordered his prisoner, Hannah, to follow. Bluffing their way past the guards, they fled to Tomasz's country home. Once there, he rejoined the

Nazi resistance fighters, promising a quick return. But the war kept the lovers separated, with Tomasz believing that Hannah died and Hannah uncertain of his fate.

Flashing forward in time to New York, Hannah contacts the Red Cross after watching the interview and learns that Tomasz is alive, still residing in a small Polish village. More than thirty years have elapsed, yet the pull of the past is irresistible. Needing to put those memories to rest, Hannah reassures her husband of her love before leaving for Poland, saying, "The Tomasz I remember risked his life for love ... Now that he's alive, it set me free." Arriving at his village, a nervous Hannah gets off the bus, while an anxious Tomasz paces by his car. Suddenly, the former lovers spot each other, speaking only with their eyes—just like the first time so many years ago.

Emotional Memories and Traumatic Memories

Hannah's vivid wartime memories are emotional and traumatic. She holds emotional memories of loving Tomasz and traumatic memories of seeing fellow prisoners beaten or killed. The terrible memories of Holocaust survivors would not have surprised William James, the philosopher and psychologist who wrote in 1890, "some events are so emotional as to leave a scar on the cerebral tissue."[1] *Traumatic memories* involve highly stressful events that produce extreme negative emotions after a person experiences or witnesses life-threatening harm—as in a physical assault, rape, robbery, auto accident, unexpected death of a loved one, or natural disaster. These memories always involve emotional events, but not all emotional events produce traumatic memories. Whereas traumatic memories are always negative, *emotional memories* can be positive or negative and usually involve relationships with other people. If negative, they are less stressful than traumatic memories.[2]

A hotly contested issue is whether emotional memories and traumatic memories are distinct types of autobiographical memory with different characteristics. Describing this divisive topic, psychologist Richard McNally put it this way:

> Some experts believe that rape, combat, and other horrific experiences are engraved on the mind, never to be forgotten. Others believe that the mind protects itself by banishing traumatic memories from awareness, making it difficult for many people to remember their worst experiences until many years later.[3]

At the heart of this issue are questions about how memory works. Does emotion typically enhance our ability to remember or make it more difficult to remember accurately? Are traumatic memories special, being recalled better or possibly worse than other memories? Can traumatic memories be forgotten and later recalled, and, if so, are special mental processes needed to explain their forgetting? I will attempt to answer these questions by research and films showing characters who are deeply troubled by persistent memories.

Memory, Emotion, and Stress

How does emotion affect memory? Studies have generally shown that emotional arousal enhances memory. Recall of words, pictures, and stories is higher for items with positive or negative affect than items that are affectively neutral.[4] Psychologists Larry Cahill and James McGaugh, for example, showed people a series of slides depicting a mother taking her young son to a hospital. For some people, an emotionally arousing narration accompanied the slides that described the boy as badly hurt in an accident with surgeons trying to save his life; for others, the same slides were paired with a neutral narration describing a routine disaster drill. Examined two weeks later, people who previously received the arousing narration remembered more of the story and accompanying slides than people who had the neutral narration. Better memory was associated with emotional arousal.[5]

How might emotional arousal influence memory? One possibility is that it alters the manner in which we pay attention. With increases in arousal, we focus attention more narrowly, zeroing in on an event's central elements to produce better memory for those elements than other more peripheral details.[6] People witnessing an armed robbery, for instance, often focus on a robber's gun—a central element—instead of the perpetrator's clothing—a peripheral detail. This phenomenon, called *weapon focus,* shows how arousal and memory are connected.[7] Although emotionally arousing events are more likely to be remembered than emotionless events, an event's positive or negative affect also matters for memory. Negative affect, says psychologist Elizabeth Kensinger, is particularly likely to enhance memory for the intrinsic, central elements of an event.[8]

Neurologically, the brain structure most important for processing emotion is the *amygdala,* located in each temporal lobe. Taking its name from the Greek word for *almond* because of its shape, the amygdala registers the emotional significance of incoming sensory information. With links to the brain's hippocampus and frontal lobes, this structure helps in quickly deciding if an event is sufficiently menacing to warrant a fight or flight reaction. During an emotional experience, the amygdala enhances memory by modulating the effects of stress-related hormones—cortisol and epinephrine—saying, in effect, to the rest of the brain: "Hey, this is important. Remember it!"[9] When that experience is later recalled, the amygdala and related structures are similarly reactivated and the emotion is reexperienced.[10]

Memory for the details of emotional events can vary due to differences in how the brain processes positive and negative affect. We are more apt to remember where we sat during a funeral than a wedding, for instance, because negative events receive more sensory processing, including their sights, sounds, and smells, than positive events, allowing for more vivid, if not necessarily more accurate, negative memories.[11] Even in conversations, we remember criticism better than praise.

There are obvious survival benefits for remembering negative experiences if faced with similar situations in the future. Recalling a ghastly car crash while approaching a busy intersection, for example, provides a valuable reminder for caution. But if that memory persists every time you get behind the wheel, this constant reminder could become stressful. Quickly learned and remembered, highly negative experiences can interfere with daily living if they produce persistent stress.

Symptoms of Long-Term Stress

According to the American Psychiatric Association's *Diagnostic and Statistical Manual of Mental Disorders*, persistent stress resulting from a traumatic experience can lead to *post-traumatic stress disorder* or *PTSD*. This disorder is defined by negative symptoms that last for more than six months following exposure to a traumatic event involving actual or possible death, serious injury, or sexual violation. People with PTSD experience four clusters of negative symptoms: *intrusion symptoms* such as involuntary traumatic recalls, dreams of the trauma, and flashbacks in which they relive the event; *avoidance and numbing symptoms* including trying to avoid distressing thoughts and the people and places associated with the trauma; *changes in cognition and mood symptoms* such as blaming themselves for the event, losing interest in ongoing activities, and having difficulty concentrating or sleeping; and *arousal symptoms* including being hypervigilant or easily startled. Many PTSD sufferers also report substance abuse problems, numbing their painful memories by drinking or taking drugs.[12]

Following a traumatic event, it is not unusual to experience some of these negative symptoms temporarily. Many people, for example, were stressed in the weeks after the terrorist attacks on September 11, 2001, including first responders who witnessed the horrific sight of people leaping from the burning towers.[13] But most people showing symptoms of stress after trauma do not develop PTSD. It is estimated that more than half of all people will experience a traumatic event in their lifetime, but only a small percentage of them, roughly 7 percent, will get PTSD.[14] This disorder received intense coverage after the 1960s as veterans of the Vietnam War returned home.

War-Related Post-Traumatic Stress Disorder

Military combat is a risk factor associated with PTSD. Archival medical records show evidence of PTSD-like symptoms in soldiers who fought in the American Civil War, and these negative symptoms continue to be observed up through the recent conflicts in Iraq and Afghanistan.[15] When tested several decades after serving in Vietnam, combat veterans with PTSD recalled fewer positive memories to cue words such as *loyal* and *kind* than combat veterans without PTSD. The former soldiers who wore their medals and fatigues to testing seemed fixated on the time of the war, recalling many of their personal memories from that period.[16] To determine which factors were associated with PTSD,

Vietnam veterans were surveyed on their degree of combat exposure, prewar psychological vulnerability, and involvement in harming civilians or prisoners. To varying degrees, PTSD was associated with all of these factors, with severe combat exposure leaving veterans most vulnerable to this disorder.[17]

Numerous films involving the Vietnam War, including Michael Cimino's *The Deer Hunter,* Francis Ford Coppola's *Apocalypse Now,* and Stanley Kubrick's *Full Metal Jacket,* depict wartime trauma, but they do not show the consequences of these experiences on a soldier as fully as Oliver Stone's *Born on the Fourth of July*. Based on Ron Kovic's autobiography and featuring Tom Cruise, Frank Whaley, Willem Dafoe, and Kyra Sedgwick, this film follows a small-town boy from childhood to adulthood—to hell and back.

"He's my Yankee Doodle boy."
Born on the Fourth of July

Patriotism was never questioned in Massapequa, Long Island, in 1960. Gathered around their TV sets, families watched a youthful president promising to "pay any price, bear any burden ... to ensure the survival and success of liberty." Young Ronnie Kovic took President Kennedy's words to heart, later enlisting in the US Marines after turning eighteen. Many of his childhood friends followed, even though few of them understood what was happening on the other side of the world.

A gung-ho marine in Vietnam, Ron never questions the war until one engagement shatters his body and beliefs. Ordered to destroy an enemy village, Ron's squad lets loose a deadly barrage, killing or badly wounding everyone inside. When the dust settles, the marines discover that they mistakenly slaughtered women and children. Pandemonium quickly ensues as nearby Vietcong soldiers attack and the marines scatter for cover. In the fog of combat, with everyone yelling and shooting, Ron fires at a figure running toward him. Tragically, he kills nineteen-year-old Billy Webster, before he himself is seriously wounded, leaving him paralyzed in a wheelchair for life.

Back home in the late 1960s, with America deeply divided on the war, Ron serves as the grand marshal for his town's July 4th parade. But stopping midsentence in his speech, flinching as firecrackers go off, he is overcome by a wartime flashback—hearing gun battles raging in his mind. As the hushed crowd looks on, his boyhood friend Tim rescues him, another soldier back from Vietnam. Both men are suffering from PTSD. Ron self-medicates with alcohol, trying to avoid thinking of his horrific experience, while Tim has his own bad memories. Asked how he handles them, he replies, "Mostly, I do a lot of drugs ... You do it any way you know how."

Coming home drunk most nights, Ron clashes with his devoutly religious mother, culminating one night with both of them screaming at each other. Finally telling her of the village massacre, he pleads, "Thou shalt not kill, Mom. Thou shalt not kill women and children ... Remember. Isn't that how you taught us?" Unable to believe, unable to accept what has become of her son, she orders him out.

Rock bottom is a Mexican resort town, full of wounded veterans, where Ron connects with Charlie, another vet in wheelchair, drinking his way to oblivion. Eventually taking their frustration out on each other, the men fight on the side of a desert highway, knocking each other to the ground. Lying in the sand, Ron says, "I had a town once Charlie. I had a mother and father. Things that made sense. You remember things that made sense, things you could count on before we all got so lost?" His tears silently falling, Charlie remembers. Ron's spiritual rehabilitation has begun.

Bound for Georgia to visit Billy Wilson's parents, Ron stops at Billy's gravesite before going to his home. The parents were told that their son died valiantly serving his country, and Ron must tell them the truth. Trembling as the family quietly looks on, Ron describes how their boy really died.

> I remember the day he was killed ... People were yelling ... firing at anything, and that was when it happened ... God, this is very difficult for me to say (sobbing) ... I think I was the one who killed your son that night. I was the one.

The room is silent after this confession, until Mrs. Wilson offers absolution by saying, "We understand, Ron."

Radically changed by war, Ron will march with fellow antiwar veterans at the 1972 Republican Presidential Convention and speak at the 1976 Democratic Convention, affirming his deeply held moral beliefs.[18] In his autobiography, his words echo those written long ago by William James:

> I wanted people to ... really know what it meant to be in a war. ... To kill another human being ... is something that never leaves you. ... That scar and memory and sorrow will be with you forever.[19]

Non-War-Related Post-Traumatic Stress Disorder

Combat exposure is not the only factor associated with PTSD. Trauma can have serious consequences following noncombat experiences. Elizabeth Ebaugh went food shopping in 1986 and was abducted by a knife-wielding man as she loaded groceries into her car. After driving around for hours, the man took Ebaugh to a motel where he raped her, before forcing her back into her car. Stopping by a deserted bridge at 2 a.m., the man bound her wrists and threw her into the frigid water below. Somehow, she swam to safety and was rescued by a friendly truck driver.

As a result of this night of terror, Ebaugh suffered long-lasting PTSD. She had recurring frightening memories, avoided going near the grocery store and bridge, and experienced a heightened state of arousal. Overcoming her fears took years, even with the close support of family and friends. But today, shopping at that store and driving over that bridge are no longer scary. Like many people who experience PTSD, she eventually bounced back.[20]

Non-war-related PTSD can also occur if you harm others by accident or neglect. Intense feelings of remorse are especially common for survivors of motor vehicle accidents in which someone has died, a traumatic experience that can trigger PTSD.[21] In Jonathan Demme's *Rachel Getting Married*, featuring Anne Hathaway, Rosemarie DeWitt, Bill Irwin, and Debra Winger, the death of a child from a car crash has long-lasting consequences for the driver, as well as other members of the family.

"Have you seen Mom?"
Rachel Getting Married

Rachel is marrying Sydney, but everyone is a bit on edge because Rachel's sister, Kym, is to be part of the wedding. Home temporarily from a drug rehabilitation center, Kym quickly senses her family's mistrust—from her father, who denies her use of his car, her sister, who chooses a friend as her maid of honor, and her mother, who remains emotionally distant. The family's only wish is for things to go smoothly this weekend.

In and out of rehab since adolescence, Kym made a terrible mistake and has been coping with this bad memory for years. Her problem stems from her teenage drug addiction. Driving with her little brother, Ethan, Kym was sixteen and high on drugs when she lost control of the car and drove it off a bridge into a river. She alone survived the plunge. Reminders of Ethan are everywhere at home, producing intrusive memories and filling her with deep remorse. Desperate to make amends, Kym discusses her rehab at the prenuptial dinner, but this public display does not sit well with Rachel. Later that evening, the sisters have a heated argument, with Kym storming off in her father's car in search of her mother, producing a confrontation long in the making. Following Ethan's death and the turmoil that Kym created, her mother and father divorced.

Agitated over her fight with Rachel and questioning why her mother left her in charge of Ethan, Kym projects her guilt over his death on to her mother. Quickly, their emotions get out of hand.

> Kym (yelling): You knew. ... I was a crazy drug addict ... Why would you leave me in charge of him?
> Mother (raising her voice): Because you were good with him.
> Kym (yelling louder): Mom, Mom, why would you leave ... a drug addict to watch your son?
> Mother (now screaming): You were not supposed to kill him!

Enraged, mother and daughter scuffle, trading slaps before Kym races off in her father's car, crashing it headlong in the woods. Uninjured, save for a black eye, returning home in a tow truck, Kym goes to the one person who might help her, despite the many heartaches she has caused. Finally seeing Kym's frail, childlike state clearly, Rachel lovingly bathes and dresses her sister, before the wedding begins.

Amid the daylong festivities, hoping for reconciliation, Kym cannot catch her mother alone. Her only chance comes later that evening when, leaving early, her mother offers a strained farewell. Warily avoiding eye contact with Kym, she says, "You take care, all right?" And then she drives off. Only before heading back to rehab the next morning does Kym get the affection she craves—an embrace from her older sister. At least one relationship is mended.

Memory, Trauma, and Forgetting

In *Rachel Getting Married,* Kym is tormented by the memory of accidentally killing her brother. Could she possibly ever forget it? Alternatively, what if something terrible, such as physical or sexual abuse, happened to her as a little girl? Could she ever forget that? Your answers to these hypothetical questions depend on your beliefs about memory.

Understanding the fate of traumatic memories is complicated because there are competing views about how memory functions. Some mental health clinicians who examine trauma in the context of psychotherapy hold that traumatic memories are special. Operating differently than other memories, highly distressing memories can be forgotten when a special mechanism called *repression* deletes these uncomfortable thoughts from conscious awareness. These bad memories may resurface years later in the course of psychotherapy or come forth spontaneously in response to environmental reminders.[22] Many memory researchers disagree, arguing that while people might well wish to forget traumatic events, these experiences are typically remembered and no special forgetting mechanism is necessary.[23]

Further complicating this matter, both sides differ on the best evidence for deciding the issue. Clinicians who believe that traumatic memories are special hold that memories reported in psychotherapy are more relevant to understanding how trauma affects individuals than laboratory studies of memory, whereas memory researchers—knowing that autobiographical recollections can be untrustworthy at times—rely on carefully controlled experimental research.

These conflicting views came to a head in the late 1980s with news reports of individuals claiming that they were sexually abused as children, forgot the abuse, and then remembered their traumatic experiences years later, often during psychotherapy. As families were torn apart by lawsuits alleging abuse by a family member, these cases became a national concern, providing a pressing need to understand trauma's effect on memory.[24]

What is *childhood sexual abuse* or *CSA*? Legally, the definition varies across states and countries, based on different definitions of childhood and ages of consent, but all include a range of inappropriate behaviors between an older person and a child that are designed to sexually arouse the adult. From hospital and police records, the reality of CSA, including instances of exposure, touching, and penetration, is unquestioned, with prevalence rates in the United States ranging from 19 to 28 percent for females and 9 to 16 percent for males.[25] Precise statistics are hard to obtain as a result of varying definitions of abuse and the potential for underreporting.

Abused children can experience a range of psychological troubles, including depression, anxiety, and guilt, along with behavioral problems such as truancy and delinquency. Their recovery varies with their age and understanding, whether they confided in a trusted adult who believed them, and how quickly they reported it. Children who disclose their abuse quickly tend to be less traumatized than others who keep it a secret. When a trusted adult commits the abuse, the impact can be severe and long lasting, triggering PTSD.[26] Several films, including Gregg Araki's *Mysterious Skin* and Amy Berg's documentary *Deliver Us from Evil,* show the deep-rooted consequences of CSA committed by predator adults, as does Clint Eastwood's brooding film,

Mystic River. Featuring Sean Penn, Tim Robbins, and Kevin Bacon, it reveals one man's persistent distress from the abuse that he experienced as a boy.

"The thing is, it's like vampires, once it's in you, it stays."
Mystic River

Mystic River tells the tragic tale of three childhood pals—Jimmy Markum, Sean Devine, and Dave Boyle—growing up in a working-class neighborhood of Boston. Playing in front of their houses one day, the boys scratch their names in the wet cement of a newly poured sidewalk. As Dave is writing his name, a car pulls up and a man flashing a badge jumps out. Claiming to be a cop and berating the boys for damaging city property, the man orders Dave into the car, where a priest, riding up front, silently smiles. As Jimmy and Sean look on, Dave is whisked away and sexually abused by both men for four days—at one point begging his abusers, "Please, no more. Please"—before he manages to escape. Embarrassed and afraid, this eleven-year-old boy keeps this terrible experience a secret, but he is forever changed.

Twenty-five years later, the former best friends are connected only by their past. Sean is a homicide detective, Jimmy is an ex-con who runs a convenience store, and Dave is a blue-collar worker who, like Jimmy, never left the old neighborhood. Their lives become reconnected after Jimmy's nineteen-year-old daughter, Katie, is murdered—the same night that Dave arrives home wounded and bleeding, telling his wife, Celeste, that a mugger slashed him. When Celeste hears about Katie's murder the next day, she begins doubting his story. Dave was one of the last people to see Katie alive at a local bar around 1 a.m., and the police, led by Sean, view him as a prime suspect.

Celeste grows increasingly suspicious after finding Dave watching a horror movie, talking cryptically about vampires infecting their victims and his own boyhood trauma.

Dave: I never told anyone before ... Dave was the boy who escaped from the wolves.
Celeste: What are you talking about?
Dave: They took me for a four-day ride. And ... did they have their fun ... I don't know who came out of that cellar, but it sure ... wasn't Dave.

Not understanding his terrible memory, Celeste conveys her suspicions to Jimmy, who had promised his dead daughter revenge: "I'm going to kill him, Katie. I'm going to find him ... and I'm going to kill him." When Sean arrests Katie's killers a few days later—two boys who pulled a deadly prank—it is too late. Jimmy and his thugs, the Savage brothers, have already murdered Dave.

Back on the street curb where the story began, Sean tells a half-drunk Jimmy of the arrests and that Dave is wanted for questioning about the murder of a pedophile. Adding that Dave has not come home, Sean knows about Jimmy's violent past and that his boyhood friend will not return.

Sean: When was the last time you saw Dave, Jimmy?
Jimmy: It was twenty-five years ago. Going down this street in the back of that car.

Silently, both men look down the road where Dave was driven away, their memory of that day as enduring as their names still etched in the pavement.

Remembering Trauma

The depiction of Dave as someone who always remembered his nightmarish experience is consistent with an American

Psychological Association report written by clinicians and memory researchers saying "most people who were sexually abused as children remember all or part of what happened to them although they may not fully understand or disclose it."[27] Given that abuse is usually remembered, could it be forgotten and later recalled? Could a "recovered" memory be authentic or might it be false, fostered inadvertently by a suggestive therapy technique? Twenty years ago, the answers were unclear.

Ethical and humanitarian considerations have prevented memory researchers from tackling this issue directly. Researchers, for example, could never expose anyone to a real-life trauma to see if it could be forgotten and subsequently remembered. Instead, researchers have addressed this issue by studying trauma's effect on memory through surveys and case studies and by experiments on the conditions that could give rise to *false memories*—instances, say psychologists Charles Brainerd and Valery Reyna, in which we have definite memories of events that did not actually happen.[28] This research sought to determine if traumatic memories could be forgotten, if forgotten traumatic memories could be recovered, and if recovered memories could be authentic.[29] At issue is whether the basic memory processes discovered by cognitive researchers are sufficient to apply to everyone—including people who have experienced trauma.

Can Memories of Traumatic Events Be Forgotten?

Surveys in the form of interviews or questionnaires are used to sample people's behavior and beliefs, including people's recollection of past sexual abuse. Several studies have surveyed adults with documented records of CSA to see if it is recalled. One study used hospital records of abuse to locate young adult women who had experienced forced sexual contact during childhood.

During interviews that did not mention the specific past abuse, the majority of the women recalled it, but 38 percent did not, even though they discussed other personal matters.[30] Another study used court records and found that 81 percent of the young women and men surveyed disclosed their prior abuse in interviews that did not ask about their specific cases.[31] In these studies, the older the child at the time of abuse, the more likely the abuse was recalled in the interviews. Some people never reported their documented CSA—possibly because they forgot it, perhaps because they were reluctant to discuss it, or maybe because they were too young to understand what it was when it happened—but most people reported it. Traumatic memories are usually remembered, often with a high level of vividness and sensory detail.[32]

This finding is almost certainly the case for traumatic memories involving time-extended events. When Holocaust survivors were first interviewed in the 1940s and again in the 1980s, their memories of the concentration camp, although imperfect, were highly consistent. These survivors, like Hannah in the film *Remembrance*, never forgot their living conditions in the camp, including their treatment by the guards.[33]

Can Forgotten Memories of Traumatic Events Be Recovered?

Case studies are in-depth accounts of individuals, and psychologist Jonathan Schooler has collected accounts from six people who claim that they were abused during childhood, forgot their abuse, and recovered their memory of it years later.[34] One person, identified by the initials JR, was thirty years old when he watched a movie involving sexual molestation. Later that night, he suddenly recalled being molested by his parish priest when he was eleven. Seeking corroboration, Schooler found that JR's brothers said that the priest had also approached them and the priest himself later acknowledged molesting JR.

Schooler found corroborating evidence of abuse in all six cases, but there was no way to verify their claims of forgetting, other than their statements of being shocked by suddenly remembering their earlier abuse. Their claims of prior forgetting may, in fact, be accurate, but there is no way to know for certain. Some individuals, in fact, forgot that they had talked about their abuse to others during a time when they claim to have forgotten it.[35] Can memories of abuse be forgotten and recovered? These case studies suggest that they can, but they are inconclusive.

Can "Recovered" Memories Be Authentic?

Memory researchers know that while autobiographical memory is generally accurate, it can be inaccurate at times, sometimes in surprising ways. In one study, young adults tried to remember four events from their childhood. Three events were true, obtained from a parent, and the fourth was false: "When you were five, you ran around at a wedding and knocked the punchbowl onto the lap of the bride's parents." After several days of thinking about these events, visualizing them again and again, nearly half of the people falsely remembered the wedding incident, embellishing it with lots of sensory detail, including the bride's sopping wet parents.[36] Other studies have used repeated imaginings of fictional childhood events to show that people can remember being lost in a shopping mall or being seriously attacked by an animal.[37] Similarly, when a child's photo is inserted into a photo showing a hot air balloon ride, repeated study of the doctored photo can lead to a false memory of a balloon ride that never took place. Some people even recall the view while they soared overhead.[38]

The studies of imagination-induced false memory show how we sometimes confuse imagining an event with remembering it, a type of *source monitoring error*.[39] When a fictional childhood event is suggested by a trusted person, presented in the context of real past events, and repeatedly visualized over time, a confidently held false memory can be produced. These false memory studies are relevant to remembering childhood abuse because some of the same suggestive techniques shown to produce false memories in the laboratory have been employed in psychotherapy. A 1995 survey of therapists in the United States and Great Britain found that many of them listed a wide variety of symptoms as potential indicators of childhood abuse, and 71 percent of these clinicians used various suggestive techniques to help clients recover suspected abuse memories.[40] This is a problem, says psychologist Elizabeth Loftus, because suggestions may lead to false memories.[41]

These techniques, including repeated prompting of memories, repeated imagining of events over time, and hypnosis, have produced reports of childhood abuse and, in some individuals, fantastical memories of alien abduction and satanic abuse.[42] Hypnosis, for instance, increases both accurate and false recall, while inflating people's confidence in their memory.[43] Attempting to recapture forgotten events through hypnotic age regression is especially dangerous, argues psychologist John Kihlstrom, because "it is first and foremost a product of the imagination, and any accurate memory produced is likely to be blended with a great deal of false recall."[44] These suggestive techniques, intended to uncover the source of a person's troubling symptoms, can inadvertently lead to the production of false recall—even highly fanciful memories—making their use in therapy indefensible.[45] After a rush of recovered memory cases in the 1990s, a number of people have retracted their claims and successfully pursued lawsuits against their former therapists, often individuals with limited training in clinical science.[46]

For any recollection of abuse, being right is essential if someone is to be accused, but the evidence is not always clear-cut. In Andrew Jarecki's documentary, *Capturing the Friedmans*, chaos envelops the Friedman family following charges of sexual abuse by neighborhood children, some of whom only remembered it during hypnosis.

"I still feel like I knew my father very well."
Capturing the Friedmans

Through home movies, we meet Arnold and Elaine Friedman, living comfortably in Great Neck, Long Island, in 1984, raising their three sons: David, Seth, and Jesse. To boost his schoolteacher's income, Arnold opens a computer school in his home basement, where he and his then nineteen-year-old son, Jesse, teach local children, mostly boys between eight and ten, how to use these new machines. Shortly thereafter, trouble begins after Arnold is arrested for receiving a child pornography magazine in the mail.

With his sons watching from outside, the police search the family home and discover a stash of illegal magazines hidden in Arnold's study. They also find the names of the children enrolled in his computer classes. Fearing the worst, the police start interrogating the children, seeking evidence of wrongdoing in the basement. The enraged townspeople are demanding swift action, and the detectives comply, telling some of the children in their meetings that they already know what happened. One former student recalled it this way:

> What I do remember is the detectives putting me under a lot of pressure to speak up. ... I started to tell them things I was telling myself that it's not true. ... "just say this to them in order to get them off your back."

Several children provide highly incriminating testimony, saying that Arnold and Jesse physically and sexually abused them on numerous occasions. Astonishingly, no physical evidence of abuse is produced, and no child ever complained of abuse to a parent or showed any distress when picked up after class. The reports of abuse started with the police interrogations and continued with reports from children under hypnosis in therapy. Children who did not report it were said to be in denial.

As the investigation proceeds, the family begins to crumble, with Jesse adamantly denying the charges and his father saying little, other than "Nothing happened." But Elaine is unsure about Arnold, arguing on camera with her sons:

> They wanted me just to lie, you know, and say, "he didn't do it," whether I believed it or not. And I was so angry at Arnold and what he'd done that I wouldn't do it, and I said, "well, I don't know." That is the truth. I didn't know.

While claiming innocence over these charges, Arnold later reveals that he had previously sexually abused two children of a former business partner as well as his younger brother. When this case goes to court, he pleads guilty, attempting to spare Jesse, but both men are convicted in 1988 of multiple counts of CSA. Arnold died in prison from an apparent suicide, while Jesse served thirteen years before being released and is still fighting to clear his name. Throughout this documentary, the oldest son David, the one who recorded the videos, cannot imagine any wrongdoing by his father.

> We begged him to tell us that something happened, to explain how this whole mess could have happened. ... He told us nothing happened. That's good enough for me.

But by Arnold's own admission, bad things happened in the past.

Capturing the Friedmans shows how complex cases involving childhood abuse can be. It casts doubt on the evidence incriminating Arnold and Jesse, noting how some children felt compelled to report abuse they never experienced and others remembered it only when aided by hypnosis. Yet Arnold is ultimately revealed as a pedophile, even if he is not guilty as charged. Jesse's case, revealed through his attorney, has its own puzzling twists and turns. Similar to John Patrick Shanley's film, *Doubt,* viewers of this documentary are left wondering what, if anything, really happened in the Friedman household, even with so much of it recorded in home movies.

The State of "Recovered" Memories

No psychological or neurological test can determine if a recovered abuse memory is true or false, but the context for remembering abuse turns out to be highly important. Studies have shown that it is easier to find corroborating evidence for a CSA memory if it was remembered outside of therapy than within a therapist's office.[47] Moreover, people who spontaneously recall their abuse on their own are able to verify it as often as people who always remembered their early abuse.[48] Ross Cheit, for example, awoke one night after dreaming about his former camp counselor and the next day recalled that this man had fondled him in bed. Hiring a private investigator, Cheit tracked down his abuser and got a tape-recorded confession. This memory was accurate, corroborated by Cheit and others abused by this man.[49] But absent such verification, the authenticity of any recollection is uncertain. As psychologists Henry Roediger and David Gallo warn, "Much evidence exists that even detailed, highly confident recollections can be completely erroneous, so that simple recollection should not be routinely accepted as completely accurate."[50]

In 1995, a report by clinicians and memory researchers concluded that it is possible to recover an authentic memory of abuse, although this is a rare occurrence, and it is also possible to fabricate a convincing pseudo-memory for abuse that never occurred.[51] Over the past two decades, research has shed additional light on this contentious issue. We now know that some people do not report documented childhood abuse, but most people do. We also know that people who say they forgot about their abuse and subsequently remembered it say they are shocked by their discovery. If accurate, these accounts indicate that corroborated abuse can be forgotten and recovered, although there is no way to verify whether forgetting occurred. Finally, we now know that suggestive memory techniques can produce false recollections for a variety of childhood events. When these methods are used in therapy, they have the potential to produce false recollections. A growing awareness of this troublesome issue has reduced, but not eliminated, the use of these suggestive techniques by mainstream clinicians today.[52]

We remain susceptible to various memory distortions because the act of remembering is inherently a reconstructive process. Sometimes these errors are inconsequential, as when we misremember what we wore to a party; other times they can be significant, if, say, we give faulty eyewitness testimony or mistakenly recall abuse that never occurred.[53] But, as I noted in the opening chapter, these infrequent errors are normally outweighed by remembering that is sufficiently accurate to ensure our survival and flexible enough to plan and execute future action—memory's primary adaptive function.[54]

Finally, is a special repression mechanism needed to explain any forgetting of trauma that occurs? If repression is defined as a protective mechanism that banishes disturbing thoughts from conscious awareness, its use as an explanation for forgetting cannot be verified in clinical anecdotes and it has never been compellingly demonstrated in laboratory studies.[55] Alternatively, if it is thought of as consciously trying not to think about a traumatic event, this active suppression can have the paradoxical effect of enhancing trauma recollection.[56]

Memory researchers hold that all experiences, including traumatic events, have the potential to be partially or completely forgotten, and well-documented memory processes can explain this forgetting.[57] One process, for example, involves the *principle of encoding specificity*. It holds that our remembering is enhanced when the conditions at the time of remembering match those at the time we experienced an event.[58] Recall the story of Ross Cheit. After awakening in bed from a dream about his former counselor, he recalled that this man had fondled him in bed at camp. Encoding specificity can help explain why some traumatic memories are persistently recalled and others are not. If external or internal reminders, such as movies or dreams, trigger the recall of traumatic memories, then the presence of those reminders can increase the likelihood of recalling a traumatic event. But if those cues are absent, the traumatic memory may not come to mind.[59]

Oftentimes, the problem is remembering trauma, not forgetting it. If reminders of trauma are difficult to avoid, a better approach is to change a person's interpretation of those reminders, as in Elizabeth Ebaugh's case where the shopping mall and bridge no longer prompted her memories of abuse. Too often, troubling memories persist until they are adequately treated.

Learning to Live with Troubling Memories

Whenever we think about a past event, that reactivated memory becomes temporarily labile, enabling it to be modified by new information. When recalling a motor vehicle accident, for instance, our memory of it is open to editing and updating—shaped by conversations with other witnesses, news accounts that we may have read or heard, and our current thoughts. In recalling the accident, we engage neural processes similar to those used when we first experienced it, a process called *memory reconsolidation*. Thinking of the accident from time to time keeps it firmly established in memory, but it also provides opportunities for that memory to be altered. This means that people with toxic memories may find relief by editing and updating their interpretations of traumatic events.

One treatment approach involves the use of drugs, such as propranolol, that block the action of stress hormones and dampen a person's emotional reaction after a traumatic event. When propranolol was administered to people in an emergency room and for ten days after a serious accident, they showed fewer stress symptoms three months later and weaker memories of the trauma than similar patients who received a placebo.[60] Propranolol is believed to reduce troubling symptoms and weaken memory by blocking reconsolidation. A traumatic event is less likely to be recalled and reconsolidated if it does not get an initial emotional boost from the brain's stress hormones. Long-term PTSD sufferers, however, may not benefit from drugs that block reconsolidation, says psychiatrist Roger Pitman, because well-established memories are more resistant to alteration than memories for recent experiences.[61]

An alternative approach is to use psychotherapy to alter a person's interpretation of a traumatic event. In cognitive behavioral therapy, for example, therapists attempt to change the way PTSD sufferers think about traumatic events by discussing those events with them, identifying the fearful or upsetting thoughts associated with danger, and restructuring how people remember the troubling events so that they can feel safe once again.[62] A caring psychiatrist in Robert Redford's *Ordinary People*, featuring Timothy Hutton, Donald Sutherland, Mary Tyler Moore, and Judd Hirsch, combines this therapy with other methods to show how a persistent traumatic memory can be successfully updated.

"It keeps coming. I can't stop it!"
Ordinary People

In *Ordinary People*, we experience the aftermath of a tragic death in the Jarrett family. Calvin and Beth's two teenage sons, Buck and Conrad, got caught in a squall while sailing their small boat. Buck drowned, while the younger Conrad survived, but he cannot live with his guilty memory. After attempting suicide by slashing his wrists and spending four months in a psychiatric hospital, Conrad is back home, feeling alienated from family and friends. He is tense and jittery, unable to study or eat, and the dark circles under his eyes tell us that he is having trouble sleeping. He has recurring nightmares of the boating accident ending with him clinging to the overturned boat while his brother slips under water. Suffering from PTSD, Conrad is a prisoner of this awful memory.

Wanting family life to return to normal, his father exclaims at one point, "We'd have been all right if there hadn't been any mess." He is referring to the tragic accident, but the mess applies as well to their superficial family relationships that have been messed up all along. Buck was Beth's favorite, and she has grown increasingly cold and stiff around Conrad. Calvin tries to help, but all he can do is suggest that Conrad see a psychiatrist named Berger. Reluctantly, Conrad goes, seeing Berger twice a week.

At first, their meetings are uneventful, with Conrad saying little, other than that everything is fine. Gently challenging him, Berger asks why he is seeing him if everything is all right. Conrad is awkward and nervous with Berger, but his will to survive is strong. He has already survived a boating accident and suicide attempt, and now, dealing with Berger, he is driven for change and growth. As Conrad heals, he is increasingly unable to accept the status quo at home, even as his parents try to reestablish their formerly idealized lives. For Conrad, change comes in the form of a new girlfriend and Berger, who continually challenges his interpretations of people and events.

A therapy breakthrough comes one evening when Conrad learns that a friend from the psychiatric hospital has taken her life. Badly shaken by the news, Conrad rushes out, meeting Berger at the office. Once there, Conrad finally opens up, telling Berger the sad news of his friend's suicide, before describing that awful day when his brother drowned. As Conrad remembers the scene, initially blaming himself, Berger pushes him to see it differently—as a terrible accident that was not anyone's fault.

> Conrad: Why did he let go?
> Berger: Maybe you were stronger. Did that ever occur to you?
> Conrad: You just do one wrong thing.
> Berger: Um-hmm. And what was the one wrong thing you did?
> Conrad: I hung on. I stayed with the boat.
> Berger: Exactly. Now, you can live with that. Can't you?

The climatic therapy session ends with Conrad now seeing the loss of his brother as tragic, but no longer blaming himself. When his father begins therapy with Berger, trying to make sense of his life, his mother leaves, unwilling to address the emotional gulfs that have formed between her and the men in her family. Some people are capable of change; others are terrified of it.

In closing, we might wonder what possible adaptive function Conrad's bad memory could serve. Clearly, emotional memories have value in helping us remember important experiences longer and more vividly than trivial events, a possible carryover from our ancestral past when recalling the location of food or dangerous predators was essential for survival. But what function might traumatic memories serve, especially if they trigger PTSD? Collectively, the films in this chapter suggest an answer. While all of them show how debilitating a painful memory can be, two of them—*Ordinary People* and *Born on the Fourth of July*—reveal how these memories can also promote positive growth and change. Conrad Jarrett's story is fiction; Ron Kovic's memoir is based on his real-life experiences.

Fade-Out

Emotion enhances memory for positive and negative experiences, serving an adaptive function by aiding our retention of personally important events. Traumatic memories, involving highly stressful experiences, can produce PTSD with troubling symptoms. With proper treatment, these experiences can lead to personal growth and change. Our autobiographical memory is generally reliable, but the same processes that produce true memories may occasionally generate false recollections.

7 Understanding the Reality of Amnesia

Setting the Scene

Memory loss due to amnesia has a long history in film—usually portrayed as resulting from a blow to the head or a severe emotional trauma, followed by a dramatic recovery. In real life, amnesia is a relatively rare disorder caused by multiple factors, leading to temporary or permanent impairment. Popular films, including several discussed in this chapter, perpetuate old myths by showing fictional characters that bear little resemblance to actual case studies of amnesia.

Featured Films

Desperately Seeking Susan, directed by Susan Seidelman, with Rosanna Arquette, Aidan Quinn, and Madonna (Orion Pictures, 1985, Rated PG-13).

The Music Never Stopped, directed by Jim Kohlberg, with Lou Taylor Pucci, J. K. Simmons, and Julia Ormond (Roadside Attractions, 2011, Rated PG).

The Return of the Soldier, directed by Alan Bridges, with Julie Christie, Glenda Jackson, Ann-Margaret, and Alan Bates (20th Century Fox, 1982, Rated PG).

The Majestic, directed by Frank Darabont, with Jim Carrey, Martin Landau, and Laurie Holden (Warner Bros. Pictures, 2001, Rated PG).

Primal Fear, directed by Gregory Hoblit, with Richard Gere, Edward Norton, and Laura Linney (Paramount Pictures, 1996, Rated R).

Memento, directed by Christopher Nolan, with Guy Pearce, Joe Pantoliano, and Carrie-Anne Moss (Newmarket Films, 2000, Rated R).

Where Am I?

Film characters have been losing and regaining their memory for a hundred years. After banging her head in a train wreck in 1915's *Judy Forgot,* Marie Cahill lost her memory until a second head knock restored it. Fighting off crooks in 1918's *De Luxe Annie,* Norma Talmadge forgot her past following a head blow only to have it restored by a brain operation. The origin of these mythical cures for amnesia is mystifying. Years earlier, in 1882, psychologist Théodule-Armand Ribot reported his pioneering studies of *amnesia*—meaning literally *without memory*—making it likely that silent era filmmakers knew that head injuries could lead to memory loss. However, neither a second concussion nor a brain operation has ever restored anyone's lost memories. Yet, as shown in *Desperately Seeking Susan,* Susan Seidelman's zany film with Rosanna Arquette, Aidan Quinn, and Madonna, such fanciful ideas persist. This lighthearted romantic comedy employs amnesia for entertainment, not education.

"It all started with the personals. There was this ad for Susan."

Desperately Seeking Susan

Reminiscent of the screwball comedies from the 1930s, this fast-paced comedy of errors shows a bored housewife named Roberta pouring over the personals in a tabloid's classified section. One that begins "Desperately Seeking Susan," written by someone named Jim, particularly intrigues her. Jim, a struggling musician, and Susan, a dazzling seductress, plan a rendezvous in New York's Battery Park before Jim heads out on tour. Curious, Roberta decides to watch, and later trails Susan to a secondhand clothing store where she buys Susan's trademark jacket. Unfortunately, a killer has also been stalking Susan, believing that she has a priceless pair of stolen earrings. Seeing Roberta in the jacket, the killer mistakes her for Susan. As they tussle, Roberta falls, loses her purse, and develops amnesia after banging her head.

With no identification and unable to remember her past, a bewildered Roberta is rescued by Jim's friend, a handsome young man named Dez who, believing also that she is Susan, says, "You're not at all what I expected." Now thinking that she is Susan, Roberta lands a job as a magician's assistant while her husband and killer separately pursue her in a madcap plot full of improbable coincidences. This dizzying farce about mistaken identity and amnesia concludes when a second head bonk restores Roberta's memory and she realizes that Dez is the lover she was desperately seeking.

Desperately Seeking Susan is fun to watch, but it provides little understanding of the reality of amnesia. When Roberta initially strikes her head in a fall, the impact could damage those delicate neural structures important for memory. But while a concussion can lead to memory loss, another concussion never produces memory recovery. A second concussion days or weeks after the first can produce harmful brain swelling, sometimes leading to death.[1] Curing amnesia by a second head bonk may be amusing in film, but it is wildly misleading; I will speculate on the origin of this nonsensical idea later. Regarding amnesia's usual portrayal in film, psychologist Sallie Baxendale said it best: "Memories aren't made of this."[2] It is time for some facts.

The Reality of Amnesia

Amnesia refers to the temporary or permanent loss of memory function beyond normal forgetting. When a person has difficulty learning or remembering new information, it is called *anterograde amnesia*. Operating forward in time, this form of amnesia was shown in the first chapter by Drew Barrymore's character in *50 First Dates*. Lucy could remember her distant past, but she could not form new memories after injuring her brain in a car crash. Alternatively, when a person loses memory of the past—for days, weeks, or even years—the deficit is called *retrograde amnesia*. Roberta in *Desperately Seeking Susan* showed this type of memory loss that works backward in time. Movie characters such as Lucy and Roberta portray people with pure anterograde or retrograde amnesia, but in real life, both types of impairment can occur in the same person—with some individuals showing more anterograde than retrograde loss, and others showing the reverse. No two injuries are alike, and recovery, if it occurs, varies from person to person.

People with amnesia often retain their ability to converse, recall historical facts, and acquire new skills, demonstrating that aspects of their working, semantic, and procedural memory systems are preserved.[3] What defines their amnesia is their dysfunctional episodic memory—memory for the experiences that make up their life. Individuals with anterograde amnesia have difficulty forming new episodic memories, whereas those with retrograde amnesia have trouble recalling those memories. Memory impairment following brain damage is called *organic amnesia,* often shown by anterograde and retrograde loss, whereas memory failure without precipitating brain damage is labeled *functional amnesia,* revealed by retrograde loss. Traditionally, functional amnesia has been attributed to psychological or emotional trauma.[4] It lends itself to dramatic depictions in film, with characters losing their identity and forgetting large chunks of their past, but organic amnesia occurs more frequently than functional amnesia and is better understood.

Attacking the Brain: Organic Amnesia

Organic amnesia refers to a cluster of neurologically based memory disorders brought on by a host of possible factors including concussions, strokes, infections, and tumors. A cursory overview of several brain structures important for memory will help make sense of these different disorders.

An adult brain, soft and spongy to the touch, weighs approximately three pounds and contains 100 to 120 billion nerve cells.[5] These cells form neural tracts that transmit messages throughout the brain, connecting different structures that serve basic life-support functions, such as regulating breathing and heart rate, as well as higher cognitive functions, such as generating memories and thoughts. Because damage to these structures can disrupt these various functions, the brain is enclosed in a bony skull, surrounded by cerebrospinal fluid that acts as a shock absorber for minor head bumps. But these defenses offer limited protection when the brain undergoes attack by a major concussion, infection, or change in blood flow that limits the supply of oxygen to its cells. Damage to areas buried deep inside the brain including the *thalamus* and *hypothalamus*—important for relaying sensory information and regulating the release of hormones—and the *hippocampus* and *amygdala*—necessary for forming memories and emotionally tweaking them—can result in profound amnesia.

Traumatic Brain Injury: Concussions

The term *concussion* takes its two meanings from its Latin roots. It can mean either *collision* or *violent shaking*, corresponding to two types of traumatic events that can lead to brain injury. The first is a blow to the head that is severe enough to damage the brain's soft tissue. This type of concussion can occur while falling off a bike and hitting the pavement, striking a car's dashboard in a crash, or being on the receiving end of a head-banging tackle in a football game. The second type can happen in whiplash accidents when the head is rapidly thrown back and forth, causing the brain to slam repeatedly against the inside of the skull. Whiplash concussions, caused by violent shaking, are especially damaging for infants. In each case, severe stress on the brain can injure delicate neural structures leading to all manner of problems, including memory impairment.

When a twenty-year-old man, identified only as ML, was taken to the hospital after suffering a concussion in a car crash, his neurological tests were normal, yet he was unable to recall the accident. Emotionally withdrawn and repeatedly asking the hospital staff the same questions, he could not retain new facts and was unable to recall events that occurred before the accident. Although ML had a girlfriend for six months, he could not remember her name, even though it was tattooed on his forearm. Fortunately for him, the accident was not severe enough to produce lasting damage. ML's anterograde and retrograde amnesia cleared up over time.[6]

Another young man, Adam Burns, wrote about a whiplash concussion that he suffered after being hit by a car. His recovery from anterograde amnesia was slow, but he was eventually able to describe the frustration of living with this impairment. Even simple conversations were difficult:

Friend: I just bought a car!
Adam: Wow, that's great.
Friend: So do you want to go for a drive?
Adam: But you don't have a car.
Friend: I just bought one.

Unable to remember what his friend bought, Adam was too embarrassed to ask.[7]

These cases provide glimpses into the varying effects of a cerebral concussion. Immediately following his concussion, ML showed symptoms of brain injury including poor attention, slurred or incoherent speech, disorientation, and problems of remembering—symptoms that do not typically reveal themselves in a brain scan. With Adam's more moderate damage, there were additional symptoms including persistent headache, dizziness, fatigue, irritability, and an intolerance of bright light and loud noise that lasted for weeks. Still others with severe brain damage can have seizures and lapse into a coma.[8] Unlike Roberta's brief bewilderment in *Desperately Seeking Susan*, a blow to the head sufficient to produce amnesia reveals a host of behavioral problems—including some degree of anterograde and retrograde amnesia that may clear up, but only gradually over time.

In actual cases of amnesia, retrograde memory loss is greater for events that immediately precede a concussion than events that are more distant. This phenomenon, observed by Ribot over one hundred years ago and called *Ribot's law*, suggests that older memories are more stable than recent memories, especially when compared to events that occurred just before a traumatic brain injury.[9] In fact, memory for a concussion and events that immediately preceded it is often permanently lost. The driver who received a severe concussion but survived the 1997 Paris car crash that killed Princess Diana, for instance, never remembered the events preceding the accident and was unable to tell others what happened.

In cases of limited brain injury, amnesia can clear up gradually over days, weeks, or months. Recovery often follows a backward path with the oldest, strongest memories returning before memories of more recent events.[10] Yet for some events, such as those immediately before a concussion, people will learn of these events only indirectly, based on what others have told them. These case studies suggest that the brain needs sufficient time for the memory of an event to *consolidate*, to strengthen so that it is established in long-term memory and less vulnerable to future disruption.[11] If consolidation is incomplete, the memory for the event may never become part of a person's episodic past, the way that older, established memories are.

Traumatic Brain Injury: Electric Shocks

In a 1774 pamphlet, Benjamin Franklin reported that an electrical shock to the head could produce temporary amnesia. While experimenting with electricity to see if he could store it in Leyden jars, this scientist-statesman accidently shocked himself. After taking a jolt through the head that caused him to forget the shock, Franklin wrote, "It was some Moments before I could collect my Thoughts." Trying his experiment again, this time with six men—each with his hand on another's head—Franklin found that the men fell to the floor and did not recall how they got there. Fortunately, he avoided electrocuting them by using a mild shock. Others tinkering with electricity remarked that they felt mentally elated the morning after a brief shock and wondered if electricity might be useful for treating depressed people.[12]

Today, severely depressed people who do not respond to psychotherapy or drug interventions may receive *electroconvulsive therapy* or *ECT*, consisting of a series of brief electrical charges passed through their temples while under sedation. ECT produces a loss of consciousness, followed by retrograde amnesia that may extend backward up to a year or more before gradual remission.[13] Portrayed earlier as a cruel punishment given to Jack Nicholson in Miloš Forman's film *One Flew Over the Cuckoo's Nest*, ECT is now a refined medical tool used for people suffering from major depression. It may help those prone to suicide by causing them to forget troublesome issues, thereby providing time for more conventional therapies to be used.

Traumatic Brain Injury: Strokes

Changes in blood flow to the brain can also bring forth changes in memory, dramatized in the next chapter by the film *Amour*. Although the brain accounts for only a small part of a person's body weight, its cells consume approximately 25 percent of the body's need for oxygen. A *stroke*, brought on by a blockage in blood flow or a ruptured brain artery, causes brain cells to die within minutes owing to a lack of oxygen—producing symptoms such as dizziness, weakness on one side of the body, and speech problems—that can lead to permanent brain damage and possibly death. For stroke survivors, there can be lasting physical and cognitive impairments, including profound anterograde amnesia.[14] Other times, a temporary reduction in blood flow can leave little permanent damage.

One memory disorder suspected of being related to a temporary blood flow disruption is *transient global amnesia* or *TGA*. This disorder, reports psychologist Alan Brown, most often strikes people between the ages of fifty and seventy, is sudden in its onset, and produces anterograde and retrograde amnesia that quickly resolves.[15] During a TGA attack, people will be aware of their memory problem and repeatedly ask others to explain what happened. These people know who they are, can identify family and friends, but they keep asking the same questions because they are unable to remember the answers. Brain scans during a TGA episode reveal a temporary reduction in the blood supply to the hippocampus, triggered in some cases by strenuous exercise, psychological stress, and even sexual intercourse. Following sex with her husband, Alice, age fifty-nine, became disoriented and confused. At the hospital, she kept asking her husband to explain, after which she repeatedly replied: "Let me get this straight. We had sex. I wind up in the hospital and I can't remember anything? You owe me a 30-carat diamond!"[16] For Alice and others with TGA, their amnesia rapidly clears up with most feeling fine within days.[17]

Organic amnesia can develop quickly following traumatic brain injuries and other times more slowly from acquired brain injuries due to tumors, infections, and alcoholism.

Acquired Brain Injury: Tumors

A *tumor* is a growth of abnormal cells that may be malignant or benign. Malignant brain tumors are life-threatening cancers that can spread to surrounding tissue; benign brain tumors are not cancers, but they can still injure the brain by putting pressure on neighboring tissue as they grow. In his essay "The Last Hippie," neurologist Oliver Sacks describes meeting a man in 1977 who had a benign brain tumor removed, but remained hopelessly disoriented and confused, suffering from anterograde amnesia.[18] Jim Kohlberg adapted this case study for *The Music Never Stopped,* with Lou Taylor Pucci, J. K. Simmons, and Julia Ormond.

"Gabriel, is this the song?"
The Music Never Stopped

Growing up in the 1960s, Gabe Sawyer loved rock music, especially songs by the Grateful Dead. Forming a band of his own but repeatedly clashing with his father, Gabe dropped out of high school and struck out for New York. Twenty years later, his parents found him in a hospital, unresponsive and amnesic from a brain tumor. Surgery removed the large tumor, but irreversible brain injury had already occurred. The tumor had damaged multiple areas, including Gabe's hippocampus in both temporal lobes, leaving him with anterograde amnesia. Thinking that he still lived in the 1960s, Gabe stopped making new memories after 1970.

Because of Gabe's love of music, Gabe's father sought out a professor who studied music's effect on the brain. Agreeing to meet with Gabe, Dianna Daly played him different types of music, but nothing seemed to connect. Unresponsive, Gabe sat motionless with a vacant stare. But something happened when she played the French national anthem, "La Marseillaise"—Gabe immediately perked up and smiled after hearing the opening bars, only to fall back into a stupor as the song continued to play.

Stymied by Gabe's behavior, Dianna was listening to the radio one night when she heard those same opening notes—only this time from the beginning of the Beatles' song, "All You Need Is Love." Could songs from the 1960s connect with Gabe, unlocking his memories from those days? Playing this song, Dianna brings him back, even if mentally he is back in the '60s.

Gabe: Oh, I love the Beatles. "Magical Mystery Tour." "All You Need Is Love," and "Sgt. Pepper" too.
Dianna: Can you tell me where you were born?
Gabe: White Plains, New York.
Dianna: Good. Can you tell me who the President is?
Gabe: Oh, Lyndon? ... the one who was shot?
Dianna: Wrong.

As the tune ends, the contact is lost and Gabe retreats inward with a blank stare.

But Dianna's discovery gives Gabe's parents a way to connect with their son. The once-distant father and son now spend hours listening and talking, as Bob Dylan and Cream sing in the background. Unable to grasp that his former girlfriend is married and that Jimi Hendricks and Janice Joplin are dead, Gabe can only talk about the 1960s, including his disappointment for missing a Grateful Dead concert—something his father gets to amend. In real life, this patient's anterograde amnesia never remitted, but thanks to Oliver Sacks, he experienced a Grateful Dead concert and, with much excitement, met the band backstage. But he could not hang on to this special memory—it was gone the following day.

Acquired Brain Injury: Infections

Another musician suffered even greater memory dysfunction than the patient treated by Sacks. In 1985, at the height of his career, classical musician and conductor Clive Wearing contacted *herpes encephalitis*, a viral infection that causes brain inflammation and swelling. This virus damaged his brain's frontal and temporal lobes, including his hippocampus, leaving him with complete anterograde amnesia and nearly total retrograde loss.

According to his wife Deborah, Clive is aware of his predicament and keeps a record of his thoughts. Thinking that he just woke up for the first time, his notes are depressingly repetitive: "2:10 p.m.: This time properly awake; 2:14 p.m.: This time finally awake; 2:35 p.m.: This time completely awake." With a memory span of perhaps ten seconds, Clive appears frozen in time, quickly forgetting what happened as soon as his attention is drawn elsewhere.

Yet, Clive retains some memory ability. He recognizes Deborah, although if she leaves the room briefly, he greets her return warmly, as though he has not seen her for days. He knows that he has children, but he cannot remember their names. And his semantic knowledge seems unimpaired, permitting him to talk about topics he knows. But, most surprising of all, Clive's musical memory is intact. Said Deborah,

> I picked up some music and held it open for Clive to see. He picked up the tenor lines and started to sing with me. ... He was singing. ... Clive could sit down at the organ and play with both hands on the keyboard.[19]

Clive was able to conduct his former choir, cuing different singers and choir sections, just as he had done before his amnesia. The virus that severely damaged his ability to recall old memories and make new ones had spared aspects of his semantic and procedural memory needed for a musical performance.[20] Established through extensive repetition, musical memory seems resistant to acquired brain injury, even, as I show in the next chapter, in patients with Alzheimer's disease.

Acquired Brain Injury: Alcoholism

Sometimes memory loss is self-inflicted. Prolonged alcoholism that damages the frontal lobes, thalamus, and hippocampus, for example, can lead to *Korsakoff's syndrome,* a disorder named after

Sergei Korsakoff, the Russian physician who described it in the late 1800s. These patients demonstrate anterograde and retrograde amnesia, a lack of awareness about their condition, and *confabulation* by unintentionally fabricating answers to questions to cover up their memory loss. During his hospital rounds, Oliver Sacks once looked in on William Thompson, an ex-grocer with Korsakoff's syndrome who mistook Sacks for a customer.

> Thompson: What'll it be today? Half a pound of Virginia, a nice piece of Nova?
> Sacks: And who do you think I am?
> Thompson: I took you for a customer.
> Sacks: Mr. Thompson, you are mistaken again.
> Thompson: My memory has been playing me some tricks ... What'll it be now, Nova or Virginia?[21]

Brain damage brought on by alcoholism, notes psychologist Alan Parkin, is linked to a poor diet resulting in a thiamine deficiency, a consequence of excessive alcohol consumption.[22] The resulting brain damage cannot be reversed, but further damage can be prevented by adding thiamine to a proper diet and avoiding alcohol use. Any recovery of memory function tends to be slow and incomplete.

By now, a pattern should be evident—whenever the hippocampus and related brain structures are damaged, memory impairment will occur. Depending on the extent of the brain injury, anterograde amnesia occurs, often with retrograde loss. Amnesia is never so simple as the movies would have you believe.

Lost and Found Memories: Functional Amnesia

Our memories of everyday events often fade quickly, whereas our recollections of traumatic experiences are usually well retained. Yet, paradoxically, on rare occasions, an intense emotional experience can be associated with memory failure. Called *functional amnesia,* this failure of remembering with no precipitating brain injury has historically been linked to emotional trauma—an idea that has long been popularized in film. In Alfred Hitchcock's classic *Spellbound,* for instance, Gregory Peck plays a character who cannot remember his past, owing to his long-standing guilt over the accidental death of his brother.

Functional amnesia can be preceded by a highly stressful experience, such as suffering public humiliation or experiencing the death of a close family member.[23] In cases where no brain injury is detected and trauma is unequivocal, it is assumed that the trauma caused the forgetting. However, the absence of a precipitating brain injury does not mean that memory failure is unrelated to brain changes. In some cases, prior head injuries have been reported, making it difficult to determine the precise origin of this dysfunction.[24] Still, people with functional amnesia show episodic memory loss—in forgetting what they did—and semantic memory loss—in forgetting who they are.

Specifying the relationship between trauma and memory in amnesia is problematic because traumatic events are typically remembered. Too often, as shown in the previous chapter, troubling memories persist. If, on occasion, trauma can produce amnesia, some mechanism must be responsible for forgetting. Traditionally, this forgetting has been attributed to repression, the idea that conscious access to troubling memories can be blocked from a person's awareness, making those memories available, but inaccessible. Yet, as I already noted, hard evidence for this type of self-protective forgetting is lacking.[25] At present, amnesia following a traumatic event is rare, and the psychological and neurological mechanisms underlying this retrograde impairment remain undetermined. What is known is that functional amnesia can clear up spontaneously in days or weeks, with many but not all people recovering their lost memories.[26]

Clinically, functional amnesia is described as a *dissociative disorder*, an alteration of consciousness affecting a person's memory and identity. Dissociative disorders occur infrequently, but dissociative experiences are fairly common.[27] Many of us, for example, have had the experience of daydreaming while driving along a familiar route only to realize later that we have no memory of what happened during the trip. This amnesia-like experience for the drive is normal because our attention wandered, rather than staying focused on the road. Yet, regardless of where our attention was directed, we remember important events in our life and maintain our sense of identity. For individuals with functional amnesia, their conscious experience is different. Some lose their memory for a portion of their past, others lose their identity along with their memory, and still others may create multiple identities with different memories. These alterations of consciousness are called *dissociative amnesia, dissociative fugue*, and *dissociative identity disorder.*[28]

Dissociative Amnesia

People with *dissociative amnesia* remember who they are, but they have forgotten a time in their past that is linked to a stressful event. Soldiers with no signs of head wounds, for example, have been reported to develop retrograde memory loss after experiencing dangerous combat.[29] Initially labeled *shell shock* in World War I, this memory loss served as the basis for Rebecca West's 1918 novel about a soldier coming home from the Great War who retained his identity but could not remember the last twenty years of his life. This story is dramatized in Alan Bridges's

The Return of the Soldier, featuring Julie Christie, Glenda Jackson, Ann-Margaret, and Alan Bates.

"There's a great gap in Chris's memory."
The Return of the Soldier

Leaving his wife Kitty at their English estate, Captain Chris Baldry, a middle-aged gentleman, set off for France to lead a wartime infantry brigade. The year was 1914. Two years later, a woman named Margaret travels to his estate with an urgent request. Holding a telegram written to her by Chris, she tells Kitty and Chris's cousin, Jenny, that Chris has returned to England and is lying ill in a hospital ward. Shocked by the news and the person who brought it, Kitty and Jenny set forth for London, only to find that Chris recognizes Jenny, a cousin he knew growing up, but not his wife Kitty.

Jenny: Chris?
Chris: Jenny, how did you know I was here ... What's happening?
Kitty: Chris, have you nothing to say to me? ... Don't you know me? I am your wife!
Chris: Wife? Go away! I don't have a wife! ... Margaret! Margaret!

Believing that he is still a young man, Chris sent the telegram to Margaret, trying to rekindle their former romance.

Back home, Chris recognizes his house and surrounding estate, but there is a twenty-year gap in his mind. Looking in a mirror is startling, he says, seeing an older man staring back. Reluctant to accept his condition, his wife Kitty arranges for Margaret to visit him, hoping that seeing her as a middle-aged woman will shock Chris out of his amnesia. Watching from afar, Kitty seethes as the former lovers embrace. Later speaking to his doctor, Chris admits that his wife seems familiar, but adds, "I know her as one knows a woman who's staying at the same hotel."

Determined to restore Chris's memory, even if it means returning him to war, Kitty tells Margaret how they lost their two-year-old son to illness five years ago. When Margaret relays this story to Chris, he falters briefly and turns away, marching back to a smiling Kitty—his memories of war and his son have returned.

Dissociative Fugue

Other individuals with dissociative amnesia can lose their sense of identity along with their autobiographical past. This more profound dissociation is called a *fugue* state, after the Latin *to flee*. Psychologist Daniel Schacter and his colleagues tell of a twenty-one-year-old man, identified as PN, who was found wandering the streets of Toronto with no identification and complaining of back pains. Brought to the hospital by the police, PN was unable to recall his name, address, or other personal information, although he knew that he was in Toronto, remembered the name of the Canadian Prime Minister, and recalled that he was nicknamed *Lumberjack*.

When PN's photo appeared in the newspaper, a cousin identified him. She reported that he had a strong, emotional bond with his grandfather who had died the previous week, providing a plausible link between PN's amnesia and a stressful event. PN, however, had no memory of his grandfather or attending his funeral until, watching television the following night, he observed an elaborate funeral service in the TV drama *Shogun*. Over the next several hours, PN gradually recalled losing his grandfather, followed by remembering more and more personal details. Oddly, after remembering his grandfather's death, PN forgot what happened during the time that he was amnesic.[30]

Similar to PN, a character in Frank Darabont's *The Majestic*, with Jim Carrey, Martin Landau, and Laurie Holden, recovers from a dissociative fugue after watching a movie.

> "No idea who you are or how you got here?"
> *The Majestic*

The Majestic is set in the early 1950s, shortly after World War II, when Congress called filmmakers to Washington to testify about their possible communist connections. Young writer Peter Appleton is called because he once belonged to an antiwar group in college. Worried about losing his job if he testifies, Peter binges at a local bar before running away. Hopping in his car and driving away that night, he bangs his head on a stone pier after accidently toppling his car in a river.

When a stranger spots him lying dazed on the shore the next morning, Peter's memory of the past is gone. Wiping the blood from Peter's head, the man says there is something familiar about his face. Others from the small town nearby say the same thing, including Harry Trimble who recognizes Peter as his son Luke, the war hero who has been missing in action for nine years. Even Adele, Luke's former fiancée, accepts Peter as Luke. Confused, Peter has no memory of these people or The Majestic, the town's movie theater owned by Harry that Luke used to love.

> Harry: I think you loved The Majestic even more than I did. You've got to remember that.
> Peter: I don't ... None of this means anything to me.
> Harry: It used to mean so much.
> Peter: How can it? Harry, I don't even know who I am.

Yet, charmed by the friendly folks and falling in love with Adele, Peter adopts Luke's life, even restoring the rundown movie theater to its once former glory. Life seems perfect until Peter's first film, *Sand Pirates of the Sahara,* is shown one night at the theater. Watching the film, Peter starts mouthing the dialogue as the actors say their lines on screen. Realizing that those words are his and recognizing his name on the movie poster, Peter's lost memories suddenly return, just before federal agents take him into custody, forcing him to testify before the congressional committee. But Peter is a changed man. Standing before the congressmen with a copy of the US Constitution, Peter argues for the right of free speech to the great delight of the townsfolk where, upon his return, he is accepted for the person he has become.

The Majestic is a feel-good movie, reminiscent of a Frank Capra film, but its depiction of amnesia is muddled. Peter's head injury would not produce retrograde amnesia with no anterograde impairment. Any amnesia from a serious brain injury would almost always show some anterograde loss. Neither would his lost memories return quickly following a head injury, or from watching a movie. If recovery occurred, it would be gradual and slow.

The Majestic's depiction of amnesia might actually make more sense if Peter suffered functional, not organic, amnesia. In fleeing from a stressful congressional request, Peter's memory loss and recovery seem more consistent with a dissociative fugue where memory loss may be caused by emotional stress and recovery can occur relatively quickly. But even this interpretation is inadequate. Contrary to this film's Hollywood ending, people with dissociative fugue can form new memories during their amnesia, but those memories are often lost once their amnesia is resolved, as it was in Lumberjack's case. With his shifting memories, Peter would not likely return to Adele because he would not remember her or the town. This issue becomes even more clouded when multiple identities in the same person appear to maintain different memories.

Dissociative Identity Disorder

Once known as *multiple personality disorder* and portrayed in the vintage film *The Three Faces of Eve, dissociative identity disorder* is characterized by the presence of two or more distinct identities—with different personalities and memories—that alternately control a person's behavior. These identity changes involve sudden alterations of speech, affect, and behavior that are readily observable by others. People with this disorder have gaps in their memory for personal events and often report that they were physically or sexually abused earlier in life.[31] Such reports are often difficult to verify, but one twenty-four-year-old woman—called IC and described as a world-class performer—had four different identities, evidence of abuse in her past, and could not recall a single memory before the age of ten.[32] People with dissociative identity disorder can show asymmetry of amnesia across identities—some identities are aware of the other identities, but oftentimes one identity has *interpersonality amnesia*—no conscious memory of the experiences of the others.[33] When a crime has been committed, this disorder can pose serious legal ramifications. Courts will want to know whether a claim of dissociative identity disorder is real and thereby undeserving of punishment by reason of insanity or merely a clever attempt to avoid prosecution by malingering. This dilemma serves as a plot device in Gregory Hoblit's *Primal Fear*, featuring Richard Gere, Edward Norton, and Laura Linney.

"I have spells. I lose time. I can't remember ..."

Primal Fear

Aaron Stampler, a destitute, nineteen-year-old boy from Kentucky, needs all the free legal counsel he can get, and Martin Vail, a publicity-seeking Chicago attorney, will provide it. Caught running away from a brutal murder scene, Aaron has been charged with the savage slaying of a beloved archbishop. But something is oddly amiss. When Martin visits Aaron in jail, he finds a shy, polite young man who claims that someone else murdered the archbishop in his presence. Frightened, he ran away, adding, "I blacked out. It happens to me at times. When I woke up I was covered in blood."

When a psychologist examines Aaron, she discovers that he can abruptly change his identity—switching from the meek, stuttering Aaron to the belligerent, violent Roy. The psychologist tells Martin that she believes that Aaron has multiple personality disorder, brought on by a physically abusive father. Initially skeptical, Martin observes the identity change firsthand after accusing Aaron of lying. Seeking a motive for the murder, Martin discovers that the archbishop provided Aaron and others with shelter in return for taping them performing sexual acts. Humiliated, Aaron's alter slew the archbishop.

Needing to convince a jury of Aaron's disorder, Martin instigates an identity switch when Aaron, testifying on the witness stand, changes to Roy and nearly chokes Janet Venable, the prosecuting attorney who challenges his story. This gambit pays off, leading to a quick dismissal of the case and allowing Aaron to escape prosecution. Only later, while telling Aaron that he will be remanded for treatment at a mental hospital, does Martin learn the truth:

> Aaron: Mr. Vail? Will you tell Miss Venable I'm sorry. Tell her I hope her neck is OK.

Martin: What? You told me you don't remember. You black out ... So how do you know about her neck?
Roy: Well, good for you, Marty. I'm glad you figured it out.

Until the end, *Primal Fear* provides a chilling portrait of an individual with dissociative identity disorder, with one identity having amnesia for the existence of another. In plotting his revenge against the archbishop, Aaron fooled everyone in his case. In real life, when claims of amnesia for a crime are made, malingering defendants sometimes overplay their hand, creating an atypical simulation by forgetting too much. Given a seven-digit memory span task, for instance, actual amnesia victims will remember the digits, whereas malingerers, thinking that all aspects of memory must be impaired, recall only a few numbers. Feigning amnesia accurately is not as easy as it may seem.[34]

Memory's Most Studied Mind

No account of amnesia is complete without mentioning its most famous case. Known only as HM while alive, Henry Molaison died in 2008 at the age of eighty-two. When his death was first reported, my students sent me emails, alerting me to his passing. They were always fascinated by his remarkable tale. What made Henry so special that researchers studied him for decades? The answer requires us to go back in time to 1953.

Experiencing minor epileptic seizures during childhood, Henry worsened as he aged. By the age of twenty-seven, he was having major seizures that could not be controlled by anti-convulsive medications. Incapacitated by seizures that could produce brain damage if unchecked, unable to work or live a normal life, Henry was a candidate for brain surgery. Prior to the operation, Henry's neurosurgeon took brain-wave recordings from different areas of his brain, each time looking for a site that produced excessive electrical activity that could spread over the entire brain, precipitating a convulsive seizure. If a site were found, the surgeon could remove that precise area, providing it would not impair any basic functions such as movement or speech.

Although no site of abnormal electrical activity was detected in Henry's brain, his surgeon performed what he later called a "frankly experimental operation" by removing large portions of Henry's hippocampus, amygdala, and surrounding tissue in both temporal lobes.[35] Tragically, the critical role of the hippocampus in making episodic memories was unknown at that time. After her son woke up from this radical surgery, Henry's mother quickly realized that something terrible had happened. He could not recognize his caregivers, recall recent conversations, or even find his way to the bathroom.[36] Only later was it learned that a person would retain memory function, albeit diminished, after losing one hippocampus. Henry showed what happened when both were lost. He experienced profound anterograde amnesia for the remainder of his life. Imagine for a moment the devastating effect of this operation—you are twenty-seven-years-old and for the rest of your life you will be unable to produce any new episodic memory—no new friends, no memory for movies or books, and no memory for any places you visit. This became Henry's life.

What memory functions, if any, were spared by this operation? Henry retained his large vocabulary and extensive knowledge of world events, as well as his autobiographical memories from before his operation. But, says neuroscientist Suzanne Corkin who studied Henry for many years, his personal recollections were general, not colored with specific details. Aware of his faulty memory, Henry could not remember even recent experiences, and this episodic memory impairment made it hard for him to imagine the future.

> Suzanne: Do you know what you did yesterday?
> Henry: No, I don't.
> Suzanne: How about this morning?
> Henry: I don't even remember that.
> Suzanne: What do you think you'll do tomorrow?
> Henry: Whatever's beneficial.[37]

Yet Henry could demonstrate some new memory, providing he was not asked for conscious recollection. For example, Henry acquired several motor skills after his operation, such as learning a pathway through a pattern maze and how to fold and unfold the walker that he used after taking a tumble on ice. But when shown such devices, Henry claimed that he never used them before. He demonstrated procedural memory with no conscious awareness of this learning.[38] Similarly, Henry showed evidence of implicit memory priming by completing word stems with uncommon words that he had recently been shown, rather than their more common completions—*clay* for *CLA_* rather than *clap*—but when asked to recall any recently shown words, he said that he was unable to remember.[39] Henry also acquired some surprising bits of semantic knowledge after his operation—knowing of Elvis and astronauts and a presidential assassination—possibly from the sheer repetition of watching television and reading magazines numerous times.[40] But absent any ability to produce new episodic memories, Henry could not get by on his own. He required constant supervisory care, first at home with his parents, then in a nursing home.

Assessing Henry's legacy, Corkin wrote, "Henry participated in a period of incredible change and advancement in our understanding of the brain, although he was unable to remember any of it."[41] Talking with writer Philip Hilts, Henry touched on his legacy this way when asked why he had trouble remembering:

> Henry: Well ... possibly I had an operation. And somehow the memory is gone.
> Phillip: Is that worrisome?
> Henry: Well, it isn't worrisome in a way, to me, because I know that if they ever performed an operation on me, they'd learn from it. It would help others.[42]

His good nature and generous spirit should give us pause.

One film provides a fascinating glimpse into what Henry's experience may have been like. In portraying a man with profound anterograde amnesia, Christopher Nolan's *Memento,* with Guy Pearce, Joe Pantoliano, and Carrie-Anne Moss, simulates this memory impairment in viewers.

"Now, where was I?"
Memento

In telling two stories at once, this dark tale of revenge provides viewers with a puzzle to solve. There is a primary story involving Leonard Shelby's hunt for his wife's killer—shot in color and told backward from end to beginning—and a secondary story about how Sammy Jankis unknowingly killed his wife—shot in black-and-white and told from beginning to end. By alternating the scenes of each story, we learn about Sammy from Leonard, an untrustworthy narrator with anterograde amnesia, and we learn about Leonard by watching his story unfold in reverse. This reverse order, says Nolan, puts viewers inside Leonard's head, sharing his amnesia and confusion.[43]

Leonard's life changed one night when two intruders assaulted and murdered his wife. Awakened by the scuffle in the bathroom, Leonard rushed to his wife's aid, killed one intruder, but was beaten by the other, suffering brain damage that left him with anterograde amnesia. That, says Leonard, is the last thing that he remembers. Seeking revenge for his wife's death, Leonard is aided by a corrupt cop named Teddy and a drug dealer's girlfriend named Natalie. Both take advantage of Leonard's amnesia by having him eliminate unsavory characters. Even Burt, a motel clerk, takes advantage of Leonard by renting him several rooms at the same time, knowing that he will not remember. As they talk, Leonard tries to explain his condition:

> Leonard: Since my injury I can't make new memories. Everything fades ... I don't even know if I've met you before ... I've told you this before, haven't I?
> Burt: I don't mean to mess with you but it's so weird. You don't remember me at all?
> Leonard: No!

To retain new experiences, Leonard writes detailed notes, takes Polaroid pictures, and has cryptic reminders, such as *Remember Sammy Jankis,* tattooed on his body—external substitutes for his damaged memory.

Alone in his motel room, Leonard narrates Sammy's story. As an insurance investigator, Leonard met Sammy when his diabetic wife filed a medical claim for Sammy's anterograde amnesia. Leonard denied them coverage because he found no medical reason for Sammy's condition, but Sammy's wife suspected her husband of faking and gave him the ultimate memory test. Knowing that Sammy would never harm her, she had him administer her insulin shots every few minutes, thinking he would surely stop. But unable to remember the injections,

Sammy kept repeating them until his wife slipped into a coma and died.

Later, Teddy tells Leonard a different version, saying that Leonard has confused his life with Sammy's, adding that Leonard's wife was diabetic and Sammy was not married. Which version is correct? Did Sammy kill his wife with an insulin overdose or is this how Leonard's wife really died? Leonard's memory may be correct, or it could be a fabricated memory designed to cover up his culpability and memory loss. Does Leonard have organic amnesia from a concussion or functional amnesia linked to the killing his wife? The film provides brief clues, but they may be nothing more than red herrings. The question of whether Sammy's story is actually Leonard's story may be unanswerable.[44] Because of his amnesia, Sammy could not have told Leonard the details of his wife's death, and Leonard would not have remembered killing his wife by an insulin overdose.

In simulating amnesia by cleverly putting viewers inside Leonard's confused head, *Memento* is an irresistible amnesia film, even though its portrayal is unrealistic. If Leonard suffered a major concussion, the last thing that he would remember would probably not be his wife's death. He would likely experience retrograde amnesia, at least for recent events. Even more egregious, Leonard frequently demonstrates new memory after his wife's death, in spite of his anterograde amnesia. While in bed with Natalie, for example, he recalls their earlier conversation, locates a pen, and writes on her photo: "She has also lost someone. She will help you out of pity." He remembers the Jaguar that he took from her boyfriend and even where he parks it each night. Leonard's remembering is needed to keep the film moving, but it passes unnoticed because it is so easy to take memory for granted—something that Henry Molaison could never do.

Remembering Again!

For all the unhappy cases of unremitting amnesia, sometimes memory function recovers, even in cases of extreme memory loss. In her book *I Forgot to Remember,* Su Meck provides an inspiring example. This wife and mother of three suffered virtually complete retrograde amnesia in her twenties when a ceiling fan fell on her head. When she woke up in a hospital bed, she could no longer read or write, do simple arithmetic, or tell time. She had to relearn how to brush her hair and hold a fork correctly, relearning that occurred slowly. Said Su, "I learned that I had parents and a lot of brothers and sisters."[45]

After experiencing losses of episodic, semantic, and procedural memory, Su relearned everything she once knew and more, often using her children to teach her—learning reading and spelling and handwriting along with her kids as they progressed through school. "It was quite a few years," she said, "before I realized that the word alleluia wasn't alligator, and amen wasn't a man, and let us pray wasn't lettuce rain."[46] Later, when relatives suggested that she try taking some college classes, Su was fearful but persisted, graduating a few years ago. Su's memories from before her accident have not returned, her retrograde amnesia apparently permanent, but she was able to use her remaining ability to make new memories and tell others her remarkable story.

Why one person recovers and another does not remains a mystery, but memory recovery from brain injury does occur, although, notes neuropsychologist Barbara Wilson, it typically requires a lengthy rehabilitation to merely reduce the impairment.[47] Why then would any film suggest that recovery from amnesia only takes a second knock on the head? Of course, films need not reflect life, and comedies are meant to amuse.

But where did this idea come from? One possibility is that filmmakers have confused functional and organic amnesia, either ignoring or not knowing their difference. Roberta in *Desperately Seeking Susan*, for example, shows only retrograde amnesia after her concussion, followed by a rapid recovery after a second head bonk—the type of memory loss and recovery more associated with functional than organic amnesia. If reexperiencing an emotional trauma can lead to memory recovery, then reexperiencing a traumatic blow might do the same—especially if amnesia is erroneously thought of as simply forgetting the past.

Fade-Out

Amnesia is far more complex than its usual portrayal in film, where confused characters wander around in a daze, followed by a quick recovery. It can involve anterograde or retrograde impairment, it can be temporary or permanent, and it has different loss and recovery characteristics related to its organic or functional origin. It makes for a dramatic story, especially in cases of recovery, but for every Su Meck, there are others like Clive Wearing and Henry Molaison whose lives have been permanently disrupted by their memory dysfunction.

8 Senior Moments, Forgetfulness, and Dementia

Setting the Scene

Seniors know that memory changes with age and that few of these changes are desirable. Some age-related changes are normal, whereas others—such as Alzheimer's disease—are pathological, brought on by brain dysfunction. Although the possibility of losing memory is worrisome for older people, research on lifestyle and successful aging suggests ways to help maintain cognitive function.

Featured Films

Cocoon, directed by Ron Howard, with Don Ameche, Wilford Brimley, Hume Cronyn, Maureen Stapleton, Jessica Tandy, and Gwen Verdon (20th Century Fox, 1985, Rated PG-13).

On Golden Pond, directed by Mark Rydell, with Katherine Hepburn, Henry Fonda, and Jane Fonda (Universal Pictures, 1981, Rated PG).

Amour, directed by Michael Haneke, with Jean-Louis Trintignant, Emmanuelle Riva, and Isabelle Huppert (Sony Pictures Classics, 2013, French with English subtitles, Rated PG-13).

Away from Her, directed by Sarah Polley, with Julie Christie, Gordon Pinsent, and Olympia Dukakis (Lionsgate Films, 2006, Rated PG-13).

The Bucket List, directed by Rob Reiner, with Jack Nicholson and Morgan Freeman (Warner Bros. Pictures, 2007, Rated PG-13).

"My Memory Isn't What It Used to Be"

In meeting people at social gatherings, I often have a predictable conversation. Upon hearing that I study memory, young people ask about my work, while older people comment on their forgetfulness. Everyone knows that memory changes over time and that a failing memory is an early warning sign of dementia. Surveys show that seniors believe that their memory has gotten worse over the past five years, with Alzheimer's disease a major health concern.[1] Sensitive about their forgetfulness, older people wonder whether their inability to think of a particular word or remember someone's name might be a symptom of impending dementia. As a memory researcher, I have these same embarrassing experiences—asking my wife to hand me "that thing" in the kitchen, meaning the potholder, or greeting an acquaintance with a friendly "Hi, good to see you!" masking my momentary blanking on a name.

These memory lapses, colloquially called *senior moments* when they occur in the elderly, often take place when we are tired, distracted, or under stress, and they are frustrating at any age.[2] President Jimmy Carter, for example, reportedly sent a suit to the dry cleaners, forgetting that it contained information needed for the nuclear launch codes. During an interview, Rod Stewart once forgot the name of his first love, the woman who inspired his signature song, "Maggie May." And when reminded that he needed to land his seaplane on water instead of a runway, Greek General George Metaxas apologized for his forgetfulness and landed safely on water, before opening the door and falling into the sea.[3] We see these anecdotes as amusing examples of absent-mindedness in younger adults, but view them more alarmingly when they happen to seniors.

What changes in memory occur normally with advancing years and what changes signal a possible descent into dementia? This is what my gray-haired friends are wondering when they tell me that their memory is not as good as it once was. Any understanding of memory and aging must distinguish between normal and pathological changes, while recognizing the hopes and fears of seniors—illustrated with dignity and tenderness in this chapter's films. Surprisingly, Ron Howard's science fiction film *Cocoon*, featuring an ensemble cast of older actors—including Don Ameche, Wilford Brimley, Hume Cronyn, Maureen Stapleton, Jessica Tandy, and Gwen Verdon—provides a realistic glimpse of the challenges confronting seniors.

"Where we're going, we won't get sick. We won't get any older."

Cocoon

Living in a Florida retirement center, three old-timers—Art Selwyn, Ben Luckett, and Joe Finley—have grudgingly accepted their age-diminished abilities. Their lives are dull and uneventful until one day, seeking a change, they enter an abandoned estate and discover an unused swimming pool. Soaking in the curiously warm water, the men feel their aches and pains washing away. Surprised, Art exclaims, "I feel tremendous! I am ready to take on the world!" The water, containing several large pods at the bottom, is oddly restorative. In no time at all, these swimming buddies are doing backflips and cannonballs, their minds again sharp and alert. Though they look the same, the water has physically and mentally rejuvenated them, even arousing long-forgotten sensual desires.

The pods at the bottom of the pool hold dormant extraterrestrials left behind by friendly aliens and kept alive by a life force in the pool, the same life force that has reenergized the swimmers. When the mother ship later arrives to return the aliens to their home planet, their leader offers the men and their wives a tantalizing offer:

> You people seem to want what we've got. Well, we have room for you. ... You would be students of course, but you'd also be teachers. And the new civilizations you would be traveling to would be unlike anything you've ever seen before.

Mindful of their mortality, the couples face a dilemma. If they stay, they will grow increasingly old and frail, and Joe already has terminal cancer. If they go, they will never be ill or die, but they will never see their family and friends again. For Ben and Mary, this means saying goodbye to their daughter and grandson. Mary has qualms about leaving, but Ben wants to go. Unafraid of dying, his only fear is losing Mary after sharing a lifetime of memories together. Will they stay or go? Each couple has a day to decide.

The films in this chapter are love stories that highlight the various mental changes that older adults can experience—troubling cognitive changes that challenge those afflicted along with their caring partners. Might these age-related changes be stopped or reversed? This ancient quest, epitomized by the fountain of youth pool in *Cocoon*, underlies those late night TV infomercials with their promises of restoring youthful vitality. Seniors worried about losing memory are especially vulnerable to spurious claims about improving memory. These worries are real, and while I will focus on memory dysfunction, these problems are not inevitable—most people do not undergo dementia. I will also describe important discoveries on maintaining memory in old age—increasingly vital research, as older adults are now the fastest-growing segment of the world's population.[4]

What Lies Ahead

Change, argued the Greek philosophers Plato and Heraclitus, is a constant of nature; all of us are subject to change. The physical changes that occur during maturation are readily apparent to all. Less obvious, but no less real, are age-related changes in memory. Although our memory systems do not decline across the board, studies show that older adults—typically retired individuals between ages sixty and eighty—do experience more difficulty than younger adults—often college students around age twenty—on various memory tasks, whether recalling names, conversations, prose passages, bridge hands, or buildings along familiar routes.[5] Yet, while seniors as a group show diminished memory, individual differences exist with some seniors remembering as well as or better than their younger counterparts.

At what age does memory begin to decline? When the same adults are tested periodically over a span of many years, called *longitudinal* testing, they show a decline after age sixty. When adults of different ages are tested at the same time, called *cross-sectional* testing, a decline starts between ages forty and fifty.[6] While there is no set age, the ability to recall experiences begins slipping after the middle age years in cultures around the world.[7] This universal memory decline is linked, in part, to neurological changes in the brain.

The Aging Brain

As we age, the brain shrinks in size, but contrary to popular belief, we do not lose thousands of brain cells each day.[8] Instead, the reductions in brain volume and weight in healthy older adults are due to neural shrinkage, not cell death, and the overall reduction in brain size is relatively modest at approximately 2 to 5 percent per decade of adulthood.[9] Shrinkage occurs in the frontal lobes—important for making plans and focusing attention—and the hippocampus—critical for establishing new long-term memory. In older adults, volume loss in the prefrontal cortex—the front portion of the frontal lobes—is associated with problems in inhibiting competing stimuli, such as tuning out extraneous conversations, while loss in the hippocampus makes it harder to form new memories, such as learning a new route to a grandchild's home.

Important for emotional processing, the amygdala shows little atrophy with age. When younger and older adults view emotionally positive scenes, such as a child playing on a beach, their amygdala activation is similar, but when looking at negative scenes, such as a couple in the hospital, a difference emerges—older people show less activation than younger adults.[10] Older people will also remember positive scenes generally as well as younger adults, while demonstrating poorer memory for negative scenes. Psychologist Mara Mather and her colleagues suspect that seniors limit their focus on negative experiences because they perceive them as a poor use of their remaining time.[11]

At the brain's cellular level, there is a reduction in the synaptic connections between neurons that conduct electrical signals, a loss of the myelin sheathing surrounding neurons that slows the transmission of those signals, and a decrease in the supply of chemical messengers, called neurotransmitters, that facilitate neural communication.[12]

All of these brain changes are normal, and they will affect some cognitive functions more than others. After studying a lengthy word list, for example, older people will typically recall fewer words than younger adults. But if people are asked to circle the words they previously studied from a set of old and new words, recognition is similar for older and younger adults.[13] This finding indicates that old age does not produce a global memory dysfunction. Widespread impairment only occurs when there are pathological changes in the brain. Neuroscientists have estimated that the brain's normal slow loss of neural connections would require a healthy adult to live to age 130 before experiencing any Alzheimer's-like memory loss.[14]

Mechanisms of Change

Different mechanisms underlie the normally occurring age-related changes in memory. These mechanisms include a general slowing of cognition and a declining ability to block extraneous stimulation. Social-cultural factors in the form of negative stereotypes can also impact memory function.

Processing speed refers to the time it takes to perform a mental task. Just as older adults no longer run as fast as they did in their youth, they cannot process information as quickly as before.[15] When asked, for example, whether pairs of letter groups, such as *MGN—MGB,* are the same or different, older and younger adults are equally accurate, but older people make slower decisions.[16] Older adults are generally slower than younger people at learning complex tasks and retrieving information from memory.

Older adults also have a harder time than younger adults in eliminating distractions while trying to focus their attention. This ability, call *inhibition,* declines in later years, making it more difficult, for instance, for older people to concentrate on reading a book while someone is talking on a cell phone nearby than when reading in a quiet room. In any task that engages working memory—our currently held thoughts—inhibition helps to block irrelevant information, such as that nearby cell phone conversation, from entering conscious awareness. Younger adults are better at inhibiting these distractions than older adults.[17]

Finally, negative stereotypes about memory and aging, prevalent in Western cultures, can subtly prompt seniors to approach new learning passively, not performing as well as they might. When seniors, for example, were flashed a list of positive or negative associates of aging—words such as *guidance, wise,* and *sage* or *senile, confused,* and *decrepit*—too briefly to be consciously seen before they studied unrelated lists of words or pictures, the seniors exposed to the positive associates recalled more items than those shown the negative associates.[18] Memory for words or pictures declines with age around the world, but this decline was found to be greater in the United States than China, a country that traditionally has held a generally positive attitude about aging.[19]

All of these factors contribute to changes in memory with age, changes that affect our working, episodic, semantic, and procedural memory systems in different ways. Depending on the system, memory can get worse, stay the same, or even improve in later years.

Aging Memory Systems

Working memory focuses on the present, enabling us to think about a limited amount of information for a short period of time. For instance, after hearing this short list of words—*dog, cow, pig, owl, ape, cat*—older adults will recall almost as many words as younger adults in perfect order. But if the task is made more complex by asking for the words to be recalled in alphabetical order, age matters—older adults will do more poorly than younger adults.[20] Mentally re-arranging the words alphabetically requires intense concentration, and tasks that make heavy demands on working memory show an age-related decline, likely due to a slowing of cognition and a reduced ability to inhibit extraneous thoughts before forgetting inevitably occurs.

Episodic memory allows us to remember past experiences for future use. Whether testing people of different age groups or the same individuals over time, numerous studies show that episodic memory declines with age.[21] Recalling past events, as well as remembering when and where to perform future actions—such as keeping a doctor's appointment—are linked to the frontal and temporal lobes, including the hippocampus, areas of the brain that can show shrinkage with increased age.

Semantic memory, the factual knowledge that we have acquired for future use, generally holds up well, contributing to the wisdom frequently ascribed to seniors.[22] General knowledge can decrease after age seventy-five, whereas vocabulary knowledge, often greater for older than younger adults, can remain stable to age ninety.[23] One aspect of semantic memory that does plague seniors is the momentary blocking on names. Frustrated recently, I could picture the face of an actor—visualizing him in the films *Big*, *Sleepless in Seattle*, and *Forrest Gump*, but I was blocking on his name that seemed on the tip of my tongue. Only when I turned my attention elsewhere, thereby inhibiting an incorrect name that stuck in my mind—Forrest Gump—did the blocking evaporate and the correct name—Tom Hanks, of course—suddenly pop into awareness. Given their increased difficulty inhibiting distractions, seniors are prone to this problem after amassing a lifetime of names in semantic memory.[24]

Finally, procedural memory enables us to perform skilled actions, such as hitting a golf ball in the center of a fairway or manually shifting gears in a sports car. With the slowing of movement and cognition in old age, seniors are not as adept as young people at acquiring new motor skills, especially those involving complex movements that engage both working and episodic memory. Whether learning to type on a keyboard or playing a musical score on a piano, older adults can learn new skills, but they require more practice than younger adults before those skills are mastered.[25]

Our memory systems do change with age. Episodic memory gradually declines, working memory and procedural memory decline for complex, not simple tasks, and semantic memory remains stable or improves, except for general knowledge and name recall that diminish in later years. If these declines become pronounced, they warn of possible pathological changes that may have already begun.

A Troubling Turn: Mild Cognitive Impairment

For 10 to 20 percent of people over age sixty-five who still live independent lives, the first sign of trouble is when they or a close associate realize that their episodic memory has noticeably slipped.[26] Everyday forgetfulness increases with more frequent misplacing of keys, losing track of bills or appointments, and having difficulty shopping or preparing a meal. Following a neurological checkup, a diagnosis of *mild cognitive impairment* or *MCI* may be warranted for otherwise functioning seniors who complain of memory problems and score below average on standard memory tests.[27]

The diagnosis of MCI, more common for men than women, is based on a clinical judgment following a battery of tests in which a person's scores are compared to those of unimpaired people of the same age and educational level. Repeating the tests after several months can show if a person's memory problem is remitting, stable, or growing worse. Tests, such as the Montreal Cognitive Assessment or MoCA, measure verbal, visual, and semantic memory with tasks that include immediate and delayed recall of words, drawing the face of a clock showing a specific time such as ten past eleven, copying a drawing of a three-dimensional cube, counting backward from 100 by 7s, repeating complex sentences, and generating members of a conceptual category, such as naming different types of animals or vegetables in a short interval.[28] A diagnosis also includes input from family members about a person's behavior at home to help rule out common mental or physical factors, including depression or illness that could impair the memory test results.

Some seniors with symptoms of mild cognitive impairment find that their memory difficulties remit over time, owing to the alleviation of mental or physical problems that hampered their test performance. Others show little change if, for instance, their difficulties stem from a minor stroke.[29] But for many with MCI symptoms, their memory difficulties worsen, advancing over time to dementia. This progression has led to the growing belief that MCI is a predementia phase of Alzheimer's disease, with approximately 10 to 12 percent of people diagnosed with MCI developing this disease each year.[30] For these people, treating their forgetfulness with drugs has shown limited success because drugs cannot reverse the pathological changes already underway in the brain, changes I will describe along with Alzheimer's disease.[31]

Several films, including Jake Schreier's *Robot & Frank* and Alexander Payne's *Nebraska*, depict older men dealing with mild cognitive impairment, but none more affectingly than Mark Rydell's *On Golden Pond*, about a retired professor with a declining memory, featuring Katharine Hepburn, Henry Fonda, and Jane Fonda.

"You're my knight in shining armor. Don't you forget it."
On Golden Pond

Married for nearly fifty years and still finding each other fascinating, Norman and Ethel Thayer spend their summers at a tranquil lake called Golden Pond. Sarcastic and grumpy, Norman is turning eighty and growing frustrated and frightened over his mental decline. Ethel, still feisty and full of energy, is sixty-nine and needs all the compassion and cheer she can muster to offset her husband's gloomy outlook, especially after one of his frequent comments about dying: "You know, Norman, you really are the sweetest man in the world, but I'm the only one who knows it."

Puttering around the cottage, Norman's forgetfulness is readily apparent. Checking to see if the phone still works, he is unable to tell the operator his number—"It has a 9 in it, that's all I know." An early family photo leaves him puzzled over the people in the picture, and when Ethel hands him a pail for picking strawberries along the old town road, Norman wanders off and gets lost in the nearby woods. Shaken and confused, Norman finally stumbles back home with an empty pail, too embarrassed to tell Ethel what happened. When she asks why he returned without any berries, Norman lies, saying that he got hungry and ate them. Only later does he admit the truth.

Norman: "You want to know why I came back so fast? ... I went a little ways into the woods. There was nothing familiar ... Scared me half to death."

Ethel: "You're safe, you old poop ... After lunch ... we'll take ourselves to the old town road ... And you'll remember it all."

That evening, to celebrate Norman's eightieth birthday, his daughter Chelsea arrives with her fiancé and his young son, Billy. Chelsea has a strong bond with her mom, but she and Norman have never been close. When she asks her mother if Norman is remembering better, he overhears and calls out, "Come on, Billy. I'll show you the bathroom, if I can remember where it is." Norman is failing, but his tongue is as sharp as ever. After Chelsea and her fiancé leave, Billy stays behind, and soon he and Norman become best friends, fishing each day and playing Parcheesi at night. Spotting the Thayers' canoe, Billy excitedly exclaims, "A canoe! Just like the Indians used." Norman, still possessing an ample wit, replies, "Actually, the Indians used a different grade of aluminum." Yet, while out fishing together, Norman occasionally slips, referring to Billy as Chelsea.

When Chelsea returns and sees the affectionate bond that has formed between Norman and Billy, a bond she always longed for herself, she has a heart-to-heart talk with her mother who encourages her to break the chill by speaking with her father, adding, "Your father does care. Deeply. I know he'd walk through fire for me, and he'd do the same for you. And if you can't see that, then you're not looking close enough." Finally reminiscing with Norman, talking about their earlier summers on the lake, Chelsea recalls her many failed attempts at doing a backflip. When her father, a medal-winning collegiate swimmer, replies—"Yeah. I do remember that"—Chelsea promptly swims out to their float, completing that unfinished business from her youth. As the leaves turn their autumn colors, a father with a failing memory and a daughter with a fresh memory have finally made peace with their past.

First You Lose Memory: Irreversible Dementia

Dementia, from the Latin *without mind,* involves the loss of intellectual functions, including memory, judgment, and abstract thinking, brought on by disease or trauma to the brain in a previously unimpaired person. Most commonly found in people over age eighty, these changes exceed those produced by normal aging and can affect all aspects of a person's life, including daily living, independence, and social relations. People with dementia suffer major intellectual decline, with memory loss as an early symptom. As the dementia progresses, a person has difficulty communicating, loses orientation for time and place, and may undergo changes in personality by experiencing paranoia or other delusions.[32] Almost always irreversible, dementia remits in only about 1 percent of all cases, most often following a toxic reaction to prescription or over-the-counter medications.[33] As with cancer or heart disease, dementia is not a single disease, but a general term covering multiple medical conditions. While it can occur in Parkinson's disease, Huntington's disease, HIV infection, and traumatic brain injury, this neurocognitive disorder is most commonly observed in cases involving strokes or Alzheimer's disease.

Strokes and Vascular Dementia

Strokes, as I noted in the preceding chapter, involve a disruption in the supply of blood to the brain. When blood flow is interrupted in the brain's vascular system by a ruptured vessel or blood clot, neural cells quickly die from a lack of oxygen and other vital nutrients, leaving a person with brain damage. A brain scan can identify the type of stroke, and emergency neurosurgery might repair a broken vessel or drugs may break up a clot to minimize the damage. Once a stroke occurs, the consequences are variable, depending on the location and extent of the brain damage.

A person experiencing a stroke can show a sudden numbness or weakness on one side of the body, suffer a severe headache, or have difficulty walking or seeing.[34] The mental and physical symptoms can occur suddenly, following a major stroke, or more slowly, following a series of *infarcts*—small strokes that are called *silent strokes* because they often pass unnoticed.[35] Over time, even silent strokes can produce brain damage so extensive that everyday behavior is profoundly and irreparably impaired. Similar to Alzheimer's disease, a person with *vascular dementia* will show slowed thinking, forgetfulness, slurred speech, language problems, inappropriate emotions, depression, movement problems, and urinary incontinence as the disease progresses—no longer able to function independently or do things that formerly came easy.

No treatment can reverse the brain damage underlying vascular dementia, a disorder affecting men more often than women and associated with the same risk factors as cardiovascular disease—high blood pressure, high cholesterol, diabetes, smoking, and obesity. Protecting the brain through preventive measures that reduce the likelihood of heart problems—such as eating a healthy diet, exercising, and maintaining a proper weight—will reduce, but not eliminate, the chances of a stroke.[36]

In her book *Knocking on Heaven's Door*, writer Katy Butler describes how her mother valiantly cared for her ailing father who, collapsing on the kitchen floor, suffered a major stroke at age seventy-nine and died six years later—after vascular dementia had ravaged him mentally and physically. Years earlier, Jeffrey Butler was a Wesleyan University professor who greeted colleagues with a radiant smile. In the later stages of the disease, his daughter details how her mother would "take him to the toilet, change his diaper and lead him tottering to the couch, where he would sit mutely for hours … as he stared out the window."[37] This same type of heroic home care is movingly portrayed in *Amour*, a film involving vascular dementia, with Jean-Louis Trintignant, Emmanuelle Riva, and Isabelle Huppert. Inspired by a family event, writer and director Michael Haneke shows an elderly man doing his best to manage the suffering of the woman he loves.[38]

"Are you sure you don't remember what just happened?"
Amour

We learn in the opening scene that Anne Laurent has died. Lying in state inside her bedroom, a halo of withered flowers around her head, Anne has been dead for some time when firemen, covering their noses, break into the Paris apartment she shared with her husband, Georges. To understand this scene, the film flashes back in time, showing a long-married couple, two former music teachers now in their eighties, living comfortably together. They like to read, listen to music, and go to concerts. Early on, they attend a recital given by one of Anne's former students, but upon returning home, they discover that someone has jimmied their front door, attempting to break in. Trouble is coming into their life.

At breakfast the next morning, while carrying on about getting the door fixed, Georges notices that Anne has stopped talking. She sits motionless at the table, a blank expression frozen on her face. Not realizing that Anne is having a stroke, but knowing that something is terribly wrong, he dabs her face with a wet towel. Slowly Anne returns, completely unaware of what just happened.

> Georges: "What's the matter? Why didn't you react?"
> Anne: "To what? ... Please tell me what's wrong. What am I supposed to have done?"
> Georges: "I put this tea towel on you face, and you didn't react."
> Anne: "When ... When was it?"

Only when shown the wet spots on her blouse from the towel does Anne accept that something has happened, but she refuses Georges's request to call their physician, insisting that all is well. But all is not well. Attempting to refill her teacup, she misses the cup, pouring tea on the table. Body movement on her right side is now unstable; Anne needs medical attention.

After surgery attempts to remove an artery blockage on her brain's left side, Anne comes home in a wheelchair, fully paralyzed on her right side and increasingly dependent on Georges. Following a second stroke, her language begins deteriorating, as vascular dementia begins destroying her mind. Georges attends to her full time, declining his daughter's request that he seek help, saying, "We always coped, your mother and I." But Anne is becoming increasingly burdensome, forcing him to hire a nurse to help out three days a week.

Acutely aware of the difficult time Georges is having, Anne grows increasingly depressed, saying, "There's no point going on living. I don't want to go on." As the disease progresses, her condition deteriorates. Bedridden, no longer communicating, and needing her most basic needs met by others, Anne has one brief moment of lucidity. As Georges speaks to her softly, she grasps his hand in a gesture of understanding, and then the moment is gone forever. Burned out from trying to honor his commitment, Georges takes her hand and says, "It's all right … it's all right. I'm here … everything's fine."

Rarely occurring before age seventy, vascular dementia is the second leading cause of dementia, accounting for 10 to 20 percent of all cases in the United States and Europe and even more in Japan. Dementia's leading cause, accounting for 60 to 80 percent of all cases, is the much-feared Alzheimer's disease.[39]

The Insidious Invader: Alzheimer's Disease

President Ronald Reagan completed his second term in office in 1989. Four years later, at a party celebrating his eighty-second birthday, he rose before hundreds of guests and offered a toast to Margaret Thatcher, Britain's former prime minister. Attending the gala was Reagan biographer, Edmund Morris, who wrote, "We all froze when he toasted her twice, at length, and in exactly the same words. There was nothing we could do but give her two standing ovations … while Dutch stood obliviously smiling."[40] A year later, the former president was diagnosed with Alzheimer's disease.

Named after Alois Alzheimer, the German pathologist who first identified this disease in 1906, Alzheimer's disease currently afflicts over 5 million people in the United States, a number exceeding the populations of Connecticut and Rhode Island combined. Because of their longer lifespans and possibly hormonal changes after menopause, women are at greater risk than men.[41] Estimates of its prevalence vary, but all show that Alzheimer's increases with age from 5 to 10 percent of all people in their seventies to 25 percent or more thereafter.[42] Caring for this many people is staggering, and lacking any prevention or treatment in the near future, the prevalence of this disease is estimated to swell in the United States to nearly 14 million seniors by mid-century, a little more than a generation away.[43]

The Progression of Alzheimer's Disease

This disease was once seen as a form of dementia that was either present or absent—you had the disease or you were disease-free. More recent accounts, including the 2011 diagnostic guidelines produced jointly by the National Institute on Aging and the Alzheimer's Association—both valuable sources of online information—describe Alzheimer's as a dementia that progresses in three distinct stages.[44] Because of its insidious nature, it is hard to know when this disease begins. It shows up first with problems in remembering, and progressively worsens over a period of years. In the *preclinical stage,* a person may show no overt symptoms initially as the disease begins silently assaulting the brain. The first red flags appear in the *mild cognitive stage* with noticeable problems in attention and memory, as the disease starts attacking the working, episodic, and semantic memory systems. These cognitive deficits can appear years before a diagnosis of Alzheimer's is made.[45] People will still remember important lifetime events, but they will start misplacing possessions, have trouble managing their finances, and begin withdrawing from social activities.

Later, in the final stage of the disease, the *Alzheimer's dementia stage,* these problems worsen as people have difficulty expressing themselves, tend to repeat ideas or questions, and lose track of the date and day. Personal experiences become hard to recall, the names of family members are lost, and spouses and children are no longer recognized. Sleep patterns change and wandering occurs, with people getting lost in once-familiar places. Writer and philosopher Iris Murdoch, who authored over two dozen novels, showed these signs of dementia before dying of Alzheimer's at the age of seventy-nine. Her last novel, published four years earlier, indicated that she was already grappling with language problems by writing much shorter sentences than those found in her earlier works.[46]

Becoming bedridden and mute toward the end, people with Alzheimer's can no longer perform basic functions, needing others for eating and for using a toilet. This irreversible decline occurs rapidly for some and more slowly for others, with an average lifespan of roughly ten years following diagnosis.[47] If the disease runs its full course, death occurs by aspiration, as the brain's respiratory center eventually stops functioning, but death following pneumonia often occurs before that happens.[48]

The profound dementia observed with Alzheimer's disease is the result of macroscopic and microscopic changes in the brain. Macroscopically, the brain undergoes dramatic shrinkage as nerve cells stop functioning, lose their connections to other cells, and wither and die. Often beginning in the hippocampus, this neural loss spreads to the frontal lobes and other areas of the cerebral cortex, explaining why problems in remembering the past and imagining the future appear early in the disease, followed by difficulties with attention, language, and reasoning.[49]

Microscopically, brain cells show two significant abnormalities—*amyloid plaques* and *neurofibrillary tangles*—that are required for a definitive diagnosis of Alzheimer's and can only by confirmed at autopsy. The plaques are clusters of sticky protein fragments, called beta-amyloid, that build up between cells; the tangles are twisted strands within a cell that prevent nutrients from flowing, leading to cell death. Plaques and tangles can be present in a normal aging brain, but they occur in much greater density in people with Alzheimer's disease. It is not yet known if these degenerative cell changes cause Alzheimer's disease or are merely a by-product of it.[50]

Researchers are currently hunting for an Alzheimer's *biomarker*—an early biological sign of this disease. Testing seniors over a period of years, one study observed cell loss in the temporal lobes, including the hippocampus, five years before a diagnosis of Alzheimer's was made.[51] Other studies have detected beta-amyloid through magnetic resonance imaging or cerebrospinal fluid analysis in the brains of people who were later diagnosed with Alzheimer's.[52] People with amyloid plaques in their brain can also show these plaque deposits in their retinas, the light sensitive cells in each eye, offering the promise of a simpler, less invasive test for early detection. Because not everyone showing cell loss or amyloid develops Alzheimer's, finding a reliable biomarker that predicts this disease is essential for identifying people who might be helped by possible treatments before this irreversible dementia takes hold.

Alzheimer's Risk Factors

Finding an effective treatment for Alzheimer's requires understanding its neurological basis. One long-held view is that Alzheimer's is caused by a buildup of beta-amyloid that forms plaques in brain cells that, together with the neurofibrillary tangles—both hallmarks of this disease—results in massive cell death. Another view relates Alzheimer's to vascular dementia, as some researchers suggest that seniors whose brains already show age-related changes may become vulnerable to Alzheimer's by impaired micro-vascular function—meaning decreased blood flow in the brain from a series of small strokes—leading to a progressive atrophy of brain structures.[53] Although Alzheimer's disease has sparked considerable research worldwide, its cause or causes remain unknown, even though many of its risk factors have been identified—those characteristics that increase its likelihood of occurrence.

The primary risk factor associated with Alzheimer's is old age, with the odds of getting this disease rising sharply in the eighth decade. Genetics can play a role, as having a variant of a gene called *APOE epsilon4 allele*—or the *e4 gene* for short—is associated with increased risk. Although Alzheimer's typically occurs in old age, it can appear much earlier. Rarely occurring before the sixth decade and accounting for only 5 to 10 percent of all cases, *early onset Alzheimer's* can strike during a person's fourth or fifth decade. Several films, including Richard Glatzer and Wash Westmoreland's *Still Alice* and Yukihiko Tsutsumi's *Memories of Tomorrow*, accurately portray this disease in its early form. Sometimes it runs in families if a genetic defect is passed on that guarantees the disease, while at other times it resembles *late onset Alzheimer's* that occurs after age sixty-five.

For late onset Alzheimer's, defective genes are implicated in some cases, but they are not the entire story. A study of identical twins—siblings with identical genetic makeup—showed that if one twin developed Alzheimer's later in life, the other twin had a 50 percent change of avoiding it.[54] Besides old age and genetics, other risk factors include being female, having a prior head injury, cardiovascular disease, depression, and neuroticism—having a worrisome personality.[55]

Taking Care of One Another

There is no cure for Alzheimer's, and treatment in the form of drugs, such as donepezil, rivastigmine, and galantamine, may reduce cognitive problems in some people in the early stages of the disease, but only temporarily. As the disease progresses, the need for caregiving increases, with more than 70 percent of Alzheimer's patients receiving family homecare.[56] But the burden on a caregiver—often an aged spouse—can be physically and emotionally exhausting, triggering depression at a rate that is two or three times higher than that of noncaregivers.[57] Many Alzheimer's sufferers will eventually require full-time nursing home care and will gradually lose all awareness of their deficit—a blessing for the patients, but not for their spouses who must carry on alone. Their troubles only continue.

Several films show people succumbing to the ravages of Alzheimer's disease, including Richard Eyre's *Iris*, Bille August's *A Song for Martin*, and Michael McGowan's *Still Mine*. But Sarah Polley's *Away from Her*, with Julie Christie, Gordon Pinsent, and Olympia Dukakis, also focuses on Alzheimer's silent victims—the loving spouses forced to cope after their lives are altered forever. I think of *Away from Her* as a sequel to *On Golden Pond*. Both show long-married couples living in a lakeside cottage, but instead of autumn leaves falling on the water, it is now the dead of winter and the lake is covered with snow. Mild cognitive impairment has given way to Alzheimer's disease.

"I think I may be beginning to disappear."
Away from Her

The Andersons are an attractive, elderly couple, living an idyllic life that is coming to an unplanned end. Grant, a retired professor, looks after his wife, Fiona, who has become alarmingly forgetful. Entertaining old friends for dinner, she offers her guests more wine, but blanking on the word *wine*, she mentally freezes while holding the bottle. Realizing her dilemma, her husband rescues her by saying, "I'll have some more wine." Another time, feeling confused, she puts a frying pan in the freezer instead of the cabinet, later telling her worried husband, "Don't worry darling. I expect I am just losing my mind."

Aware of her deteriorating memory, Fiona begins reading up on Alzheimer's disease, but Grant has a hard time accepting this fate, even after she wanders off and gets lost. Clinical testing reveals that her memory problems are severe. When asked what she would do if she smelled smoke in a theater, Fiona is stumped, finally replying that she and Grant did not often go to the movies. Understanding that her condition will only worsen and over Grant's objection, Fiona checks herself into an Alzheimer's nursing home where she will have no visitors, including Grant, for one month. Settling into her new room, Fiona makes a final request to her husband, saying, "I'd like to make love, and then I'd like you to go. Because I need to stay here and if you make it hard for me, I may cry so hard I'll never stop." Afterward, Grant drives home alone—his new life without Fiona has begun.

A month later, anxious to see his wife, Grant discovers that Fiona has taken up with another Alzheimer's patient, a man named Aubrey whom she mistakenly believes she knew long ago. They have become a twosome, with Aubrey making sketches of a youthful Fiona that now hang in her room. Emotionally crushed, Grant discusses her deteriorating condition with Kristin, the managing nurse. Asking if Fiona even knows who he is, Kristin replies, "She might not. Not today, and then tomorrow. You never know ... You'll learn not to take it so personal." For Grant, it is entirely personal.

Devastated by what he has seen, feeling abandoned and replaced, Grant slowly accepts that his life has permanently changed. While he watches Fiona sharing dinner with Aubrey, a teenager sits down beside him. At first, she thinks Fiona and Aubrey are married, but after Grant corrects her, she asks, "So, why aren't you sitting with your wife?" Grant answers, "I've learned to give her a little bit of space. She's in love with the man she's sitting with ... Just like to see her, make sure she's doing well ... I suppose it seems rather pathetic." Seeing his heart, the teen replies, "I should be so lucky."

Quickly deteriorating after Aubrey returns home, Fiona spends her days lying in bed, no longer bothering to dress or bathe. Trying to ease her gloom, Grant brings Aubrey back for a visit. Just outside of her room, Grant says that he needs a moment alone with his wife. Walking in, he finds her dressed, reading one of the books that he used to read to her. Seeing him, a once-again radiant Fiona embraces her surprised husband and says, "I seem to remember you reading this to me. You were trying to make me feel better. You tried so hard. You're a lovely man, you know. I'm a very lucky woman." The clouds have temporarily parted.

Some Surprises about Alzheimer's

Fiona's recovered memory might seem like a fanciful Hollywood ploy, but these brief periods of lucidity actually occur in Alzheimer's patients. Little is known about these surprising awakenings, other than that they are rare and temporary, lasting only minutes or hours. Musicologists working with Alzheimer's patients have reported anecdotes where music occasionally triggers these lucid interludes.[58] When songs from earlier times are played for these seniors, some patients perk up and sing, using the melody to cue the lyrics from memory. Michael Rossato-Bennett's documentary, *Alive Inside: A Story of Music and Memory*, captures some of these poignant moments when Alzheimer's patients are given iPods loaded with personalized music selected by their families. One elderly patient sings in sync with music from a 1940s band, while another dances euphorically, listening to the Beach Boys sing "I Get Around." Musical instruments can also cue an awakened performance with Alzheimer's patients playing tunes on a piano or violin from memory.[59] Music therapist Gretta Sculthorp, writing to neurologist Oliver Sacks of her experiences with dementia patients, said, "At first I thought I was providing entertainment, but now I know that what I do is act as a can-opener for people's memories."[60]

I had a similar touching experience, serving as a memory can-opener while visiting my ninety-two-year-old aunt who suffered from Alzheimer's disease. Cared for at home by her daughter, my aunt seemed lost sitting on her sofa. When I knelt down before her, I smiled and asked if she knew who I was. It was the wrong thing to ask. Of course, she did not recognize me, and my question only confused her. Quickly, I began reminiscing about my deceased mother, the sister with whom she was close. She would begin a sentence with my mother's name, and then trail off, uttering something unrelated and unintelligible. After several minutes, I gave up. I kissed her and started walking away. As I was leaving the room, she turned to her daughter and said, "That was Jackie." She used my childhood name. I had made contact after all. Whether through music or reminiscence, wonderful moments of lucidity sometimes appear. They are real and serendipitous, but cannot be forced. When they occur, they are something to cherish.

The periods of lucidity imply that some aspect of our sense of self is retained, even while Alzheimer's is wreaking havoc on the brain. Whether this sense of self is ever completely lost is impossible to know, as people in the late stage of this disease cannot be tested in any formal way. However, psychologist Stanley Klein and his colleagues have provided some insight into this issue with two intriguing case studies. The first case involved PH, an eighty-three-year-old woman diagnosed with Alzheimer's eight years earlier. PH could not recognize any recent photos of herself, but readily identified herself in photos taken during her twenties and thirties. The second case involved KR, a seventy-six-year-old woman with severe memory problems who had undergone changes in her personality owing to the disease. When given a list of personality traits and asked to select those that best described her personality, KR could do the task. But her daughter, in examining her mother's selections, said that KR had correctly identified her earlier, not her current, personality traits. In both cases, PH and KR correctly identified an aspect of their self—their appearance or personality—but it was from an earlier time in their life. Unable to update their memory, these Alzheimer's patients knew who they were, not who they have become.[61] This finding helps explain why Fiona in *Away from Her* hung those youthful sketches of herself in her room. Unlike her aged face in the mirror, the younger sketches were comforting reminders of who she was.

Finally, although Alzheimer's patients can sometimes be aggressive, becoming angry and abusive for no apparent reason, they still recognize kindness and are capable of love and affection. Surprisingly, they demonstrate enhanced *emotional contagion* by matching their emotions with others around them.[62] Yet, when emotional bonds form between patients, families can be unsettled. Like Grant in *Away from Her*, spouses of dementia patients are sometimes faced with the heart-wrenching ordeal of watching a wife or husband become attached to another patient. Former Supreme Court Justice Sandra Day O'Connor, for one, faced this dilemma while her husband was residing in a nursing home. On visits, she would find him relaxed, holding hands with his new girlfriend—another Alzheimer's patient. Rather than being distraught, she was reportedly thrilled that he was finally content.[63] Spouses can feel forsaken, until they accept the hard truth that their former partner no longer knows who they are. Patients form new relationships that are often more emotional than physical, seeking companionship with someone sharing their same confused experience.[64] As Fiona said of her new boyfriend, "He doesn't confuse me."

Alzheimer's disease can also lead to some peculiar relationships. While researching late life divorces for a new book, writer Deidre Bair came across an unusual case. A couple in their sixties decided to call it quits after being married for forty-two years. Twenty years later, after both people developed Alzheimer's and needed home supervision, their adult daughter took them in, and, to her surprise, they became the best of friends. According to Bair, "They talk about how they wish they'd met when they were younger so they could have gotten married. They're certain that they would have had a marriage that would have lasted forever."[65] And so they had.

Maintaining Memory in Old Age

Dementia is the dark side of memory that seniors fear most, even though many will only experience normal, age-related declines that, at worse, will be frustrating or socially embarrassing. To some extent, a proactive approach can offset these changes.

Everyday Memory Strategies

Regardless of age, forgetting affects all of us. We forget where we placed the car keys, forget to pick up our pills at the pharmacy, forget to buy bread at the market, and forget an appointment with the doctor. A few simple memory strategies can help minimize these problems. First, items used every day should be kept in one place. Car keys, for instance, can be kept in a bowl by the door used to enter and exit a home. Second, notes—whether Post-its, note cards, or cell phone messages—are memory extenders, external reminders of what needs to be done. Simple "To Do" lists remain popular because they work. Writing things down facilitates remembering, as does saying them out loud. You are less apt to worry about whether you locked the door after leaving home if, as you exit, you say, "I am locking the door."

Exercising the Mind and Body

The phrase *use it or lose it* has become a cliché, with many people believing that staying mentally active might prevent or delay any age-related memory declines. As a result, some seniors spend countless hours doing crossword puzzles, playing bridge, or playing brain-training games by Nintendo or Luminosity. Can these activities help maintain memory function? More time playing bridge can make you a better bridge player, but it will not make you a better poker player or produce a positive effect on memory in general. Studies of younger and older adults given extended practice on memory tasks tend to show that any benefits are limited to the specific tasks used during memory training.[66] "Focusing on just one activity," say psychologists Giles Einstein and Mark McDaniel, "does not seem to give the degree of mental exercise needed to keep the brain fit."[67]

Instead, researchers suggest that memory will be better maintained if seniors engage in an active lifestyle that incorporates a variety of everyday activities that include social interactions, hobbies, and mentally challenging activities—such as learning a second language or taking college courses online. Doing these things is important, but being intellectually engaged is not enough. Physically activity is necessary for everyone—meaning cardiovascular exercise that elevates the heart rate and increases the all-important blood flow to the brain—especially for seniors with the *e4 gene* with an increased risk for Alzheimer's disease.[68] The American Heart Association recommends that people engage in moderate aerobic exercise for thirty minutes, five times a week. Although our understanding of memory and aging remains incomplete, psychologist Carmi Schooler suggests that, at least to some degree, "using it often delays the eventuality of losing it."[69]

Successful Cognitive Aging

One of the paradoxes of dementia is that some seniors develop the plaques and tangles that characterize Alzheimer's disease but never exhibit any debilitating memory problems. A long-term study of a group of Catholic nuns who agreed to brain autopsy after death, for example, found that some nuns had these plaques and tangles, but showed only normal age-related memory declines while alive. During old age, the nuns who engaged in intellectually demanding activities better maintained their cognitive function and were less likely to develop Alzheimer's disease than nuns who were less cognitively engaged.[70]

Why are some people more susceptible to dementia than others? Clinical neuropsychologist Yaakov Stern suggests that individuals can differ in two types of reserve capacity.[71] First, individuals differ in brain size, with some people having larger brains than others, including more neural connections that permit them to withstand more brain damage before showing any memory impairment. People with larger brains and more neural connections have greater *brain reserve* than individuals with smaller brains. Second, people differ in their capacity to perform various cognitive tasks and how efficiently they perform them. People with greater capacity and efficiency have greater *cognitive reserve* than individuals with less capacity and efficiency. People with either type of reserve may be better able to withstand the progressive advance of Alzheimer's disease, dying of natural causes before any disease symptoms appear.

Recently, researchers have identified a group of people in their eighties who scored as well as or better than people several decades younger on standard memory tests. Brain scans revealed that these exceptional seniors, called *SuperAgers*, have a thicker cerebral cortex, the outermost layer of the brain needed for cognitive functioning, and fewer plaques and tangles than other seniors their age, suggesting that, for some, it is biologically possible to have an excellent memory late in life.[72]

For other seniors, the majority without an especially thick cortex, a mentally engaging lifestyle can help to maintain memory function. People who worked in complex jobs, for example, show a greater level of cognitive functioning on various memory and reasoning tasks in old age than others who were employed in less demanding jobs, even with education levels equated.[73] And when healthy seniors between the ages of sixty and ninety were given mentally challenging new activities to learn, including quilting, digital photography, or both, their ability to remember was higher three months later on various memory tests than other seniors who were engaged in either nonintellectual activities, such as playing games, watching movies, and going on field trips, or had no formal activities to perform.[74]

Memory maintenance in old age requires brain stimulation provided by learning new activities that are cognitively demanding and require sustained effort. Researchers believe that each of us has a zone of possible cognitive functioning that is influenced by our genetic background and age. But whether we function in the upper or lower portion of this zone depends on our lifestyle choices.[75] We can work to slow memory's decline and lower the risk of having Alzheimer's disease by staying intellectually active, maintaining cardiovascular fitness through aerobic exercise, reducing chronic stress, and eating a healthy diet.[76] Lifestyle is a matter of choice, whether we are young or old.

Looking Backward and Forward

On turning eighty, neurologist Oliver Sacks wrote, "I do not think of old age as an ever-grimmer time that one must somehow endure, but as a time of leisure … free to explore whatever I wish, and to bind the thoughts and feelings of a lifetime together."[77] Old age, for this wise octogenarian, is a time for both reflection and exploration, a time to use memory to look backward and forward—a theme that I have stressed throughout this book. We use old age to reminisce, to seek a coherent narrative for our lifetime experiences, and we use those experiences, as always, for imagining and planning new adventures. We only feel old, said actor John Barrymore, when regrets replace our dreams. Memory permits us to dream, to think of those things that we still want to explore.

Growing older changes how we think about age; instead of thinking about how long we have lived, we focus on how much time we might have left. For some seniors, this means compiling a list of activities to experience before dying. Rob Reiner's *The Bucket List*—as in things to do before *kicking the bucket*—helped popularize this notion. In this road film, two terminally ill seniors, played by Jack Nicholson and Morgan Freeman, embark on a final great adventure—using memory to realize their dreams, while reflecting on their lives.

"I ain't dead yet."
The Bucket List

Carter Chambers has a passion for history, once dreaming of being a history professor. But with the responsibilities of raising a family, he dropped out of college and found work as a blue-collar mechanic. His joy comes now from watching *Jeopardy* on television and answering questions faster than the contestants. Edward Cole, on the other hand, is no family man. Divorced several times, estranged from his daughter and never having met his granddaughter, his sole passion is making money by reforming hospitals and drinking Kopi Luway, the world's most expensive coffee. With nothing in common other than terminal cancer, Carter and Edward find themselves sharing the same hospital room, strange bedfellows with only months to live.

Their friendship takes root after Edward discovers that Carter has scribbled some notes on a piece of paper, before crumbling it and throwing it away. Finding it, Edward asks Carter to explain its meaning.

> Carter: My freshman philosophy professor assigned this exercise in forward thinking. We had to make a list of all the things we dreamed of doing with our lives before we ...
> Edward: Before you kick the bucket. Cutesy.

Years earlier, Carter says, he would had written, "Make a million dollars" and "First black president." Now, realizing that those are a young man's wishes, he wrote, "Witness something truly majestic," "Help a complete stranger for the good," and "Laugh until I cry." Edward, finding those choices weak, adds his own to the list, including "Kiss the most beautiful girl in the world." Telling a skeptical Carter not to worry about the cost, Edward says, "We could do this. Don't think about money. That's all I got is money." After Carter explains his quest to his doubting wife, the two men set off on their world tour—forging an unlikely friendship and living their remaining time to the fullest.

Contemplating eternity beside the pyramids in Egypt, both men begin reflecting on their life:

> Carter: The ancient Egyptians had a beautiful belief about death. When their souls reached heaven, the gods would ask them two questions. Their answers determined if they were admitted or not.
> Edward: What were they?
> Carter: Have you found joy in your life? ... Has your life brought joy to others?

Carter dies before getting to see the view from the Himalayas, and Edward still has not kissed the world's most beautiful girl. But, following Carter's urgings, he reconciles with his daughter and, kissing his granddaughter for the first time, crosses that item off the list. As the ashes of both men are left atop Mt. Everest, achieving the last wish on their list—"Witness something truly majestic"—their journey toward self-fulfillment is over. Nearing their end, they still had the courage to dream, making new memories together.

Fade-Out

All of the characters in this chapter's films have told us stories about aging, revealing, in some cases, troubling changes in memory that may lie in store with advancing age. But these films also showed us strong characters, characters like Ethel, Georges, and Grant, who provide us with important lessons about living with adversity, adding depth to our understanding of commitment and love—qualities that do not diminish with age. That is why, early on, I described each of these films as a love story and why, without memory, we cannot have stories or love.

Epilogue: Memory in Hollywood and Real Life

We spend a good deal of time watching movies. Once thought of only as entertainment, films are increasingly used—formally and informally—for education. However, caution is needed when films serve as a primary source of historical information. Films such as *Spartacus*, *Gandhi*, and *Malcolm X*, for example, offer generally accurate accounts of historical events, but in adopting a narrative structure, they also embellish their stories about these figures.[1] Similar to the autobiographical stories we tell of our past, historical stories in movies can introduce inaccuracies, and studies show that people remember the misinformation in films.[2]

In the same vein, filmmakers have taken liberties in their depictions of memory since the silent film era. For the forty films used to illustrate different memory phenomena in this book, memory has been variously portrayed—oftentimes accurately, other times not—making it risky to rely solely on movies for knowledge. Filmmakers have stories to tell and they are not obligated to present memory accurately. But it should be clear to both filmmakers and viewers when movies are bending the rules.

Bending the Rules

The idea of losing memory has long fascinated filmmakers because it gives their characters the opportunity for a life do-over, starting with a clean slate. We see Roberta Glass in *Desperately Seeking Susan* suffering amnesia from a concussion and assuming an exciting new identity until cured by a second concussion. Of course, real-life concussions never lead to memory recovery. I suspect that filmmakers have concocted fanciful tales about amnesia, in part, because stories involving true amnesic patients—people such as Clive Wearing or Henry Molaison—would be terribly depressing to watch. These memory-impaired people do not run off seeking adventure and romance.

Equally nonsensical is Christine Lucas's amnesia in Rowan Joffé's thriller *Before I Go to Sleep*. Her memory functions normally each day, but she remembers nothing from one day to the next. Inexplicably, she still possesses the wits to search for the villain who attacked her. Her nightly forgetting—like Lucy Whitmore's in *50 First Dates*—stems from a brutal knock to the head. But memory does not have a built-in timer, turning it on and off. If a head injury disrupts our ability to make lasting memories, it does so day and night.

Memento offers a more realistic glimpse of amnesia with Leonard Shelby pursuing his wife's killer—using tattoos and photos as permanent reminders because he no longer makes lasting memories. His narration about memory is often spot-on, and by telling his story backward—showing effects before causes—this film cleverly simulates Leonard's dysfunction in viewers, rendering them as memory addled as him. But, as I previously noted, even *Memento* bends the truth about amnesia—this man who cannot make new memories never forgets his way around town, where he parked his car, or that the Jaguar he drives is not his. These mundane acts of remembering—a necessary plot device for telling the film's story—are easily overlooked because we take them for granted—acts that would befuddle anyone with Leonard's condition.

Following the Rules

Yet many popular films have provided characters that depict memory phenomena accurately, even if accuracy was not the filmmaker's primary intention. Movies get memory right, for example, when characters describe their autobiographical recollections—using narrative flashbacks to show how their present circumstance was shaped by the past. When Jamal Malik in *Slumdog Millionaire* repeatedly travels back in time, explaining to skeptical police how he learned the answers to questions on a televised game show, he journeys back to his childhood—each time remembering a lifetime period, a general event, and a specific memory—the same three types of knowledge that occurs when people recall their autobiographical past.

Studies have also shown that transformative events experienced during adolescence and early adulthood produce especially long-lasting memories. The primary characters in *Stand by Me*, *Out of Africa*, and *The English Patient* all recall earlier transformative times—much like centenarian Rose Calvert in *Titanic*, who, after seeing a long-lost drawing of herself, describes in intimate detail a life-altering love affair that occurred when she was seventeen. Some memories really do last a lifetime.

Although remembering is more reconstructive than reproductive in nature, our personal memory remains a generally faithful guide to the past, especially for emotional experiences.

When former slaves recalled their time in bondage, decades after slavery ended, they provided similar accounts—faithfully reproduced in the documentary *Unchained Memories: Readings from the Slave Narratives*. Over time, details of earlier events can become fuzzy or forgotten, but the gist of such events is not lost, just as Holocaust survivors—dramatized in films such as *Schindler's List* and *Remembrance*—never forgot the horrific conditions inside the Nazi concentration camps.

Films have been especially adept at showing people haunted by their past, people who experienced or witnessed life-threatening harm. Dave Boyle in *Mystic River* always remembers what happened to him as a boy when he was kidnapped and raped by two men. Like many actual victims of childhood sexual abuse, his troubling memory persists. Also long lasting is Kym Buchman's terrible memory of crashing the family car as a teenager and killing her younger brother in *Rachel Getting Married*.

In depicting the persistence of traumatic memories, films also reveal how they can be effectively treated. When Conrad Jarrett survives a sailing accident that claimed his brother in *Ordinary People*, he suffers nightmares and debilitating stress—characteristics of post-traumatic stress disorder—until a caring psychiatrist helps him reinterpret his past and realize that his brother's drowning was not his fault. With treatment, bad memories can lead to positive growth and change.

Films have also shown that for seniors, forgetting, not persistent remembering, is the main memory worry. Aware of his increasing forgetfulness, Norman Thayer becomes terrified in *On Golden Pond* after losing his way in the woods, before stumbling back to the comforting arms of his wife Ethel. His age-related memory dysfunction—clinically termed mild cognitive impairment—can progress to dementia, leaving a person with little or no intellectual ability.

Dementia has often been accurately portrayed in film. After suffering a stroke in *Amour,* Anne Laurent rapidly succumbs to vascular dementia, even with her husband Georges attempting heroic home care. In *Away from Her,* Fiona Anderson exhibits increasing memory loss, descending into Alzheimer's disease, as does Iris Murdoch in *Iris,* and Alice Howland in *Still Alice* who experiences early-onset Alzheimer's. These films provide honest portrayals of the catastrophic effect of losing memory—both for the person afflicted and the long-suffering spouse left to carry on alone.

Occasionally, however, even films involving characters with Alzheimer's disease have exaggerated their memory ability for the sake of a story. Living in a nursing home with advanced Alzheimer's, Allie Nelson knowingly follows a story that her husband reads her each day in *The Notebook,* and Angelo Ledda carries out a final assassination in *The Memory of a Killer,* even as the disease is destroying his memory. What could be more deadly than a fictional killer with Alzheimer's?

Watching Movies with an Educated Eye

Why are some films consistent with existing memory phenomena, whereas others are merely entertaining fiction? Entertainment is important in movies and not every film genre requires realism. Reality is not required in comedies like *Desperately Seeking Susan* with its double-head-bonk view of amnesia and not expected in science fiction films such as *Total Recall*—where memories of an imaginary trip to Mars are implanted—and

Eternal Sunshine of the Spotless Mind—where memories of a former lover are erased during sleep. Viewers can put reality temporarily aside in these films and accept that memories come and go, based on each story's fictional explanation.

But dramas seem different. Viewers expect dramas to reflect real life, and filmmakers have often portrayed memory phenomena accurately when their stories involve experiences that are commonly shared. We tell our life stories using storytelling conventions, just like a narrative flashback in *The Kite Runner* or *Cinema Paradiso*; we remember a first romance long after that romance has ended, as in *Titanic*; we know what it is like to be troubled by something we saw or did, as in *Rachel Getting Married*; and many of us have witnessed firsthand the effects of dementia on relatives, as in *Amour* or *Away from Her*.

When filmmakers get memory wrong, it may be that their personal experience is lacking and their understanding is incomplete. Why would anyone imagine that amnesia caused by a concussion could be resolved by another concussion? A second concussion only makes things worse. How could memory work during the day but not overnight? Memory cannot be turned on and off like a light. Why is amnesia so often misrepresented in film? Research shows that brain-based amnesia can be temporary or permanent, often involves difficulty in forming new memories, and recovery, if it occurs, usually entails a lengthy rehabilitation. Psychologically based amnesia tends to be temporary and involves forgetting a troubling event, and reexperiencing the emotional trauma can lead to relatively rapid memory recovery. Mixing up these characteristics can lead to comical or muddled depictions of amnesia in film.

In many of the dramas that I selected for illustrating different phenomena in this book, the depictions of memory are generally consistent with scientific knowledge. Even *Memento,* with its easily overlooked errors, gets the big picture right in showing the difficulties experienced by a person with anterograde amnesia, while providing viewers with an inside glimpse of this confused state. Christopher Nolan, the writer and director of *Memento,* said of his films, "I have a faith that any audience can tell the difference between something that's consistent to rules versus something that's totally made up and anarchic."[3] This comment clearly applies to memory's depiction in the movies and why we are able to learn about memory from popular films—if we watch them with an educated eye.

Notes

1 Memory Processes and Memory Films

1. See C. N. Smith, J. C. Frascino, D. L. Kripke, P. R. McHugh, G. J. Treisman, and L. R. Squire, "Losing Memories Overnight: A Unique Form of Human Amnesia," *Neuropsychologia* 48 (2010): 2833–2840.

2. See M. Kritchevsky, J. Chang, and L. R. Squire, "Functional Amnesia: Clinical Description and Neurological Profile of 10 Cases," *Learning and Memory* 11 (2004): 213–226.

3. E. Tulving, "Introduction to Memory," in *The Cognitive Neurosciences*, ed. M. S. Gazzaniga (Cambridge, MA: MIT Press, 1995).

4. H. Ebbinghaus, *Memory: A Contribution to Experimental Psychology*, trans. H. A. Ruger (1885; Toronto: Dover, 1964).

5. J. G. Seamon, P. V. Punjabi, and E. A. Busch, "Memorizing Milton's *Paradise Lost*: A Study of a Septuagenarian Exceptional Memorizer," *Memory* 18 (2010): 498–503.

6. H. L. Roediger III, "Memory Illusions," *Journal of Memory and Language* 35 (1996): 76–100.

7. See H. L. Roediger III and K. B. McDermott, "Creating False Memories: Remembering Words not Presented in Lists," *Journal of Experimental Psychology: Learning, Memory, and Cognition* 24 (1995): 803–814; D. A. Gallo, *Associative Illusions of Memory* (New York: Psychology Press, 2006).

8. E. F. Loftus, "The Reality of Repressed Memories," *American Psychologist* 48 (1993): 518–537.

9. J. G. Seamon, M. M. Philbin, and L. G. Harrison, "Do You Remember Proposing Marriage to the Pepsi Machine? False Recollections from a Campus Walk," *Psychonomic Bulletin and Review* 13 (2006): 752–756.

10. D. A. Gallo, M. J. Roberts, and J. G. Seamon, "Remembering Words Not Presented in Lists: Can We Avoid Creating False Memories?" *Psychonomic Bulletin and Review* 4 (1997): 271–276.

11. F. C. Bartlett, *Remembering: A Study in Experimental and Social Psychology* (1932; Cambridge: Cambridge University Press, 1964).

12. U. Neisser, *Cognitive Psychology* (New York: Appleton-Century-Crofts, 1967).

13. M. K. Johnson, "Memory and Reality," *American Psychologist* 61 (2006): 760–771.

14. H. Intraub and M. Richardson, "Wide-Angle Memories of Close-Up Scenes," *Journal of Experimental Psychology: Learning, Memory, and Cognition* 15 (1989): 179–187.

15. D. L. Schacter, "Adaptive Constructive Processes and the Future of Memory," *American Psychologist* 67 (2012): 603–613.

16. See D. C. Rubin, *Memory in Oral Traditions: The Cognitive Psychology of Epic, Ballads, and Counting-Out Rhymes* (New York: Oxford University Press, 1995).

17. See R. Price, *A Palpable God* (New York: Athenaeum, 1978).

18. A. Damasio, *Self Comes to Mind: Constructing the Conscious Brain* (New York: Pantheon, 2010).

19. See K. Oatley, "The Mind's Flight Simulator," *Psychologist* 21 (2008): 1030–1032; N. K. Speer, J. R. Reynolds, K. M. Swallow, and J. M. Zacks, "Reading Stories Activates Neural Representations of Visual and Motor Experiences," *Psychological Science* 20 (2009): 989–999.

20. J. Gottschall, *The Storytelling Animal: How Stories Make Us Human* (Boston: Houghton Mifflin Harcourt, 2012).

21. See L. Cron, *Wired for Story* (Berkeley, CA: Ten Speed Press, 2012).

22. See M. Pramaggiore and T. Wallis, *Film: A Critical Introduction* (Boston, MA: Allyn & Bacon, 2011).

23. J. M. Zacks and J. P. Magliano, "Film, Narrative, and Cognitive Neuroscience," in *Art and the Senses*, ed. F. Bacci and D. Melcher (New York: Oxford University Press, 2013).

24. See, e.g., J. E. Cutting, "Event Segmentation and Seven Types of Narrative Discontinuity in Popular Movies," *Acta Psychologica* 149 (2014): 69–77; K. L. Brunick, J. E. Cutting, and J. E. DeLong, "Low-Level Features of Film: What They Are and Why We Would Be Lost without Them," in *Psychocinematics*, ed. A. Shimamura (New York: Oxford University Press, 2013).

2 Movies and the Mind's Workbench

1. P. Long, "My Brain on My Mind," *American Scholar* 79 (2010): 20–37.

2. D. L. Schacter, *The Seven Sins of Memory: How the Mind Forgets and Remembers* (Boston: Houghton Mifflin, 2001).

3. A. Baddeley, *Working Memory* (New York: Oxford University Press, 1986); A. Baddeley, "The Episodic Buffer: A New Component of Working Memory?" *Trends in Cognitive Science* 4 (2000): 417–423.

4. See S. M. Kosslyn, *Image and Brain: The Resolution of the Imagery Debate* (Cambridge, MA: MIT Press, 1994); M. J. Farah, M. J. Soso, and R. M. Dasheiff, "Visual Angle of the Mind's Eye Before and After Unilateral Occipital Lobectomy," *Journal of Experimental Psychology: Human Perception and Performance* 18 (1992): 241–246.

5. See R. J. Gerrig, *Experiencing Narrative Worlds: On the Psychological Activities of Reading* (New Haven, CT: Yale University Press, 1993); S. Umanath, A. C. Butler, and E. J. Marsh, "Positive and Negative Effects of Monitoring Popular Films for Historical Inaccuracies," *Applied Cognitive Psychology* 26 (2012): 556–567.

6. L. Cron, *Wired for Story* (Berkeley, CA: Ten Speed Press, 2012).

7. See M. Pramaggiore and T. Wallis, *Film: A Critical Introduction,* 3rd ed. (Boston: Allyn & Bacon, 2011); K. Thompson, *Storytelling in Film and Television* (Cambridge, MA: Harvard University Press, 2003).

8. "*Pulp Fiction*: The Facts (1994 Interview)," *Pulp Fiction* DVD, Disc 2, Buena Vista Home Entertainment.

9. W. James, *The Principles of Psychology* (New York: Dover, 1890).

10. See E. S. Spelke, W. Hirst, and U. Neisser, "Skills of Divided Attention," *Cognition* 4 (1976): 214–230; L. H. Shaffer, "Multiple Attention in Continuous Verbal Tasks," in *Attention and Performance V*, ed. P. M. A. Rabbitt and S. Dornic, pp. 157–167 (New York: Academic Press, 1975).

11. See J. M. Watson and D. L. Strayer, "Supertaskers: Profiles in Extraordinary Multitasking Ability," *Psychonomic Bulletin and Review* 17 (2010): 479–485.

12. P. Ratiu, I. Talos, S. Haker, D. Lieberman, and P. Everett, "The Tale of Phineas Gage, Digitally Remastered," *Journal of Neurotrauma* 21 (2004): 637–643.

13. J. M. Harlow, "Recovery from Passage of an Iron Bar through the Head," *Massachusetts Medical Society Publication* 2 (1868): 327–347.

14. See S. W. Anderson, A. Bechara, H. Damasio, D. Tranel, and A. R. Damasio, "Impairment of Social and Moral Behavior Related to Early Damage in Human Prefrontal Cortex," *Nature Neuroscience* 2 (1999): 1032–1037; A. P. Shimamura, J. S. Janowsky, and L. R. Squire, "Memory for the Temporal Order of Events in Patients with Frontal Lobe Lesions and Amnesic Patients," *Neuropsychologia* 28 (1990): 803–813.

15. A. R. Damasio, *Descartes' Error: Emotion, Reason, and the Human Brain* (New York: G. P. Putnam's Sons, 1994).

16. T. Shallice and E. K. Warrington, "Independent Functioning of Verbal Memory Stores: A Neuropsychological Study," *Quarterly Journal of Experimental Psychology* 22 (1970): 261–273.

17. G. A. Miller, "The Magical Number Seven, Plus or Minus Two: Some Limits on Our Capacity for Processing Information," *Psychological Review* 63 (1956): 81–97.

18. K. A. Ericsson, W. G. Chase, and S. Faloon, "Acquisition of a Memory Skill," *Science* 208 (1980): 1181–1182.

19. R. Schweickert and B. Boruff, "Short-Term Memory Capacity: Magic Number or Magic Spell?" *Journal of Experimental Psychology: Learning, Memory, and Cognition* 12 (1986): 419–425.

20. See L. R. Peterson and M. J. Peterson, "Short-Term Retention of Individual Verbal Items," *Journal of Experimental Psychology* 58 (1959): 193–198; M. M. Sebrechts, R. L. Marsh, and J. G. Seamon, "Secondary Memory and Very Rapid Forgetting," *Memory and Cognition* 17 (1989): 693–700.

21. S. E. Gathercole and A. D. Baddeley, "Phonological Memory Deficits in Language-Disordered Children: Is There a Causal Connection?" *Journal of Memory and Language* 29 (1990): 336–360.

22. M. Mishkin, L. G. Ungerleider, and K. A. Macko, "Object Vision and Spatial Vision: Two Cortical Pathways," *Trends in Neuroscience* 6 (1983): 414–417.

23. M. J. Farah, K. M. Hammond, D. N. Levine, and R. Calvanio, "Visual and Spatial Mental Imagery: Dissociable Systems of Representation," *Cognitive Psychology* 20 (1988): 439–462.

24. L. Piccardi, D. Nico, I. Bureca, A. Matano, and C. Guariglia, "Efficacy of Visuo-Spatial Training in Right-Brain Damaged Patients with Spatial Hemineglect and Attention Disorders," *Cortex* 42 (2006): 973–982.

25. J. Campbell, *The Hero with a Thousand Faces* (1949), in *The Collected Works of Joseph Campbell* (Novato, CA: New World Library, 2008).

26. A. Thompson, "Cameron Sets Live-Action, CG Epic for 2009," *Hollywood Reporter*, January 9, 2007.

3 Making Memories That Last

1. See J. A. Singer and P. Salovey, *The Remembered Self: Emotion and Memory in Personality* (New York: Free Press, 1993).

2. R. S. Nickerson and M. J. Adams, "Long-Term Memory for a Common Object," *Cognitive Psychology* 11 (1970): 287–307.

3. F. I. M. Craik and M. J. Watkins, "The Role of Rehearsal in Short-Term Memory," *Journal of Verbal Learning and Verbal Behavior* 12 (1973): 599–607.

4. H. Noice and T. Noice, *The Nature of Expertise in Professional Acting: A Cognitive View* (Mahwah, NJ: Erlbaum, 1997); H. Noice and T. Noice, "What Studies of Actors and Acting Can Tell Us about Memory and Cognitive Functioning," *Current Directions in Psychological Science* 15 (2006): 14–18.

5. M. Caine, *Acting in Film: An Actor's Take on Movie Making* (New York: Applause Theater Books, 1990).

6. J. Basinger, quoted in J. G. Seamon, P. V. Punjabi, and E. A. Busch, "Memorizing Milton's *Paradise Lost*: A Study of a Septuagenarian Exceptional Memorizer," *Memory* 18 (2010): 498–503.

7. T. S. Hyde and J. J. Jenkins, "Recall for Words as a Function of Semantic, Graphic, and Syntactic Orienting Tasks," *Journal of Verbal Learning and Verbal Behavior* 12 (1973): 471–480.

8. F. I. M. Craik and E. Tulving, "Depth of Processing and the Retention of Words in Episodic Memory," *Journal of Experimental Psychology: General* 104 (1975): 268–294.

9. J. S. Nairne, "Adaptive Memory: Evolutionary Constraints on Remembering," *Psychology of Learning and Motivation* 53 (2010): 1–32.

10. J. S. Nairne, S. R. Thompson, and J. N. S. Pandeirada, "Adaptive Memory: Remembering with a Stone-Age Brain," *Current Directions in Psychological Science* 17 (2008): 239–243.

11. S. B. Klein, T. E. Robertson, and A. W. Delton, "Facing the Future: Memory as an Evolved System for Planning Future Acts," *Memory and Cognition* 38 (2010): 13–22.

12. D. L. Schacter and D. R. Addis, "Constructive Memory: The Ghosts of Past and Future," *Nature* 445 (2007): 27.

13. R. L. Bruckner and D. C. Carroll, "Self-Projection and the Brain," *Trends in Cognitive Science* 11 (2007): 49–57; D. Hassabis, D. Kumaran, and E. A. Maguire, "Using Imagination to Understand the Neural Basis of Episodic Memory," *Journal of Neuroscience* 27 (2007): 14365–14374.

14. J. M. G. Williams, N. C. Ellis, C. Tyers, and H. Healy, "The Specificity of Autobiographical Memory and Imageability of the Future," *Memory and Cognition* 24 (1996): 116–125; D. Hassabis, D. Kumaran, S. D. Vann, and E. A. Maguire, "Patients with Hippocampal Lesions Cannot Imagine New Experiences," *Proceedings of the National Academy of Sciences, USA* 104 (2007): 1726–1731.

15. A. S. Brown, "The Déjà Vu Illusion," *Current Directions in Psychological Science* 13 (2004): 256–259.

16. C. J. A. Moulin, M. A. Conway, R. G. Thompson, N. James, and R. W. Jones, "Disordered Memory Awareness: Recollective Confabulation in Two Cases of Persistent Déjà Vécu," *Neuropsychologia* 43 (2005): 1362–1378.

17. J. G. Seamon, R. L. Marsh, and N. Brody, "Critical Importance of Exposure Duration for Affective Discrimination of Stimuli That Are Not Recognized," *Journal of Experimental Psychology: Learning, Memory, and Cognition* 10 (1984): 465–469; J. G. Seamon, P. C. Williams, M. J. Crowley, I. J. Kim, S. A. Langer, P. J. Orne, and D. L. Wishengrad, "The Mere Exposure Effect Is Based on Implicit Memory: Effects of Stimulus Type, Encoding Conditions, and Number of Exposures on Recognition and Affect Judgments," *Journal of Experimental Psychology: Learning, Memory, and Cognition* 21 (1995): 711–721.

18. K. W. Greve and R. M. Bauer, "Implicit Learning of New Faces in Prosopagnosia: An Application of the Mere Exposure Paradigm," *Neuropsychologia* 28 (1990): 1035–1041.

19. J. D. E. Gabrieli, D. A. Fleischman, M. M. Keane, S. L. Reminger, and F. Morrell, "Double Dissociation between Memory Systems Underlying Explicit and Implicit Memory in the Human Brain," *Psychological Science* 6 (1995): 76–82.

20. L. E. Williams and J. A. Bargh, "Experiencing Physical Warmth Promotes Interpersonal Warmth," *Science* 322 (2008): 606–607.

21. L. E. Williams and J. A. Bargh, "Keeping One's Distance: The Influence of Spatial Distance Cues on Affect and Evaluation," *Psychological Science* 19 (2008): 302–308.

22. P. G. Devine, "Stereotypes and Prejudice: Their Automatic and Controlled Components," *Journal of Personality and Social Psychology* 56 (1989): 5–18.

23. M. Shih, T. L. Pittinsky, and N. Ambady, "Stereotype Susceptibility: Identity Salience and Shifts in Quantitative Performance," *Psychological Science* 10 (1999): 302–308.

24. A. S. Baron and M. R. Banaji, "The Development of Implicit Attitudes," *Psychological Science* 17 (2006): 53–58.

25. E. G. Loftus and G. R. Loftus, "On the Permanence of Stored Information in the Human Brain," *American Psychologist* 35 (1980): 409–420.

26. W. Penfield and P. Perot, "The Brain's Record of Auditory and Visual Experience: A Final Summary and Discussion," *Brain* 86 (1963): 595–696.

27. G. R. Fink, "In Search of One's Own Past: The Neural Basis of Autobiographical Memories," *Brain* 126 (2003): 1509–1510.

28. J. C.-P. Yin, "Location, Location, Location: The Many Addresses of Memory Formation," *Proceedings of the National Academy of Sciences* 96 (1999): 9985–9986.

29. E. A. Maguire, E. R. Valentine, J. M. Wilding, and N. Kapur, "Routes to Remembering: The Brains Behind Superior Memory," *Nature Neuroscience* 6 (2002): 90–95.

30. K. A. Ericsson, "Exceptional Memorizers: Made, Not Born," *Trends in Cognitive Sciences* 7 (2003): 233–235.

31. J. Foer, *Moonwalking with Einstein: The Art and Science of Remembering Everything* (New York: Penguin Press, 2011).

32. C. P. Thompson, T. Cowan, J. Frieman, R. S. Mahadevan, and R. J. Vogl, "Rajan: A Study of a Memorist," *Journal of Memory and Language* 30 (1991): 702–724; Y. Hu, K. A. Ericsson, D. Yang, and C. Lu, "Superior Self-Paced Memorization of Digits in Spite of Normal Digit Span: The Structure of a Memorist's Skill," *Journal of Experimental Psychology: Learning, Memory, and Cognition* 35 (2009): 1426–1442.

33. D. A. Treffert and D. D. Christensen, "Inside the Mind of a Savant," *Scientific American Mind* 17 (2006): 50–55; D. A. Treffert, "The Savant Syndrome: An Extraordinary Condition," *Philosophical Transactions of the Royal Society B: Biological Sciences* 364 (2009): 1351–1357.

34. R. N. Haber, "Eidetic Images," *Scientific American* 220 (1969): 36–44.

35. E. S. Parker, L. Cahill, and J. L. McGaugh, "A Case of Unusual Autobiographical Remembering," *Neurocase* 12 (2006): 35–49.

36. B. Sparrow, J. Lui, and D. M. Wegner, "Google Effects on Memory: Cognitive Consequences of Having Information at Our Fingertips," *Science* 333 (2011): 776–778.

4 Recognizing the People We Know

1. D. Grann, "The Chameleon," *New Yorker*, August 11, 2008.

2. J. Liggett, *The Human Face* (New York: Stein & Day, 1974).

3. R. R. McCrae and P. T. Costa, "The Stability of Personality: Observations and Evaluations," *Current Directions in Psychological Science* 3 (1994): 173–175.

4. D. J. Simons and C. F. Chabris, "Gorillas in Our Midst: Sustained Inattention Blindness for Dynamic Events," *Perception* 28 (1999): 1059–1074; T. Drew, M. L.-H. Vo, and J. M. Wolfe, "The Invisible Gorilla Strikes Again: Sustained Inattention Blindness in Expert Observers," *Psychological Science OnlineFirst* (2013), July 17.

5. D. J. Simons and D. T. Levin, "Failure to Detect Changes to People during a Real-World Interaction," *Psychonomic Bulletin and Review* 5: 644–649.

6. P. Ekman, *Emotion in the Human Face,* 2nd ed. (Cambridge: Cambridge University Press, 1982).

7. V. Bruce and A. Young, "Understanding Face Recognition," *British Journal of Psychology* 77 (1986): 305–327.

8. M. M. Wilford and G. L. Wells, "Does Facial Processing Prioritize Change Detection? Change Blindness Illustrates Costs and Benefits of Holistic Processing," *Psychological Science* 21 (2011): 1611–1615.

9. K. Lah, "Is 'Minority Report' Becoming Reality?" CNN Business 360, March 11, 2010, https://artfrickus.wordpress.com/2010/12/20/is-minority-report-becoming-reality/.

10. M. Williams, "Better Face-Recognition Software," *MIT Technology Review* May 30, 2007, http://www.technologyreview.com/news/407976/better-face-recognition-software/; N. Singer, "Never Forgetting a Face," *New York Times, Sunday Business Section,* May 18, 2014.

11. N. Brewer and G. L. Wells, "Eyewitness Identification," *Current Directions in Psychological Science* 20 (2011): 24–27.

12. R. Russell, B. Duchaine, and K. Nakayama, "Super-Recognizers: People with Extraordinary Face Recognition Ability," *Psychonomic Bulletin and Review* 16 (2009): 252–257.

13. N. Kanwisher, J. McDermott, and M. M. Chun, "The Fusiform Face Area: A Module in Human Extrastriate Cortex Specialized for Face Perception," *Journal of Neuroscience* 17 (1997): 4302–4311.

14. S. Song, "Do I Know You?" *Time,* July 17, 2006, http://content.time.com/time/magazine/article/0,9171,1211572,00.html.

15. H. Sellers, *You Don't Look Like Anyone I Know* (New York: Riverhead Books, 2010).

16. M. J. Farah, *The Cognitive Neuroscience of Vision* (Oxford: Blackwell, 2000).

17. L. J. Tippett, L. A. Miller, and M. J. Farah, "Prosopamnesia: A Selective Impairment in Face Learning," *Cognitive Neuropsychology* 17 (2000): 241–255.

18. B. Duchaine, L. Germine, and K. Nakayama, "Family Resemblance: Ten Family Members with Prosopagnosia and Within-Class Object Agnosia," *Cognitive Neuropsychology* 24 (2007): 419–430.

19. O. Sacks, *The Mind's Eye* (New York: Knopf, 2010).

20. W. Hirstein, and V. S. Ramachandran, "Capgras Syndrome: A Novel Probe for Understanding the Neural Representation of the Identity and Familiarity of Persons," *Proceedings of the Royal Society: Biological Sciences* 264 (1997): 437–444; H. D. Ellis and A. W. Young, "Accounting for Delusional Misidentifications," *British Journal of Psychiatry* 157 (1990): 239–248.

5 Autobiographical Memories and Life Stories

1. M. Proust, *Remembrance of Things Past*, vol. 1: *Swann's Way*, French Pleiade edition, trans. C. K. Scott and T. Kilmartin (New York: Vintage Press, 1913–27).

2. D. P. McAdams, *The Stories We Live By: Personal Myths and the Making of the Self* (New York: Morrow, 1993).

3. Ibid.

4. C. R. Barclay and H. M. Wellman, "Accuracies and Inaccuracies in Autobiographical Memory," *Journal of Memory and Language* 25 (1986): 93–103.

5. M. A. Conway, "Autobiographical Memories and Autobiographical Knowledge," in *Remembering Our Past: Studies in Autobiographical Memory*, ed. D. C. Rubin, pp. 67–93 (Cambridge: Cambridge University Press, 1996).

6. M. A. Conway and D. C. Rubin, "The Structure of Autobiographical Memory," in *Theories of Memory*, ed. A. F. Collins, S. E. Gathercole, M. A. Conway, and P. E. Morris, pp. 103–137 (Hillsdale, NJ: Erlbaum, 1993).

7. S. Bluck, N. Alea, T. Haberman, and D. C. Rubin, "A Tale of Three Functions: The Self-Reported Uses of Autobiographical Memory," *Social Cognition* 23 (2005): 91–117.

8. G. Rizzolatti and C. Laila, "The Mirror-Neuron System," *Annual Review of Neuroscience* 27 (2004): 169–192; V. Gallese, C. Keysers, and G. Rizzolatti, "A Unifying View of the Basis of Social Cognition," *Trends in Cognitive Sciences* 8 (2004): 396–403.

9. K. Oatley, "The Mind's Flight Simulator," *Psychologist* 21 (2008): 1030–1032.

10. B. Wicker, C. Keysers, J. Plailly, J.-P. Royet, V. Gallese, and G. Rizzolatti, "Both of Us Disgusted in *My* Insula: The Common Neural Basis of Seeing and Feeling Disgust," *Neuron* 40 (2003): 655–664.

11. M. Iacoboni, *Mirroring People: The Science of Empathy and How We Connect with Others* (New York: Picador, 2008).

12. Y. Maki, S. M. J. Janssen, A. Uemiya, and M. Naka, "The Phenomenology and Temporal Distributions of Autobiographical Memories Elicited with Emotional and Neutral Cue Words," *Memory* 21 (2013): 286–300.

13. W. A. Wagenaar, "My Memory: A Study of Autobiographical Memory Over Six Years," *Cognitive Psychology* 18 (1986): 225–252; C. P. Thompson, J. J. Skowronski, S. Larsen, and A. Betz, *Autobiographical Memory: Remembering What and Remembering When* (New York: Erlbaum, 1996).

14. W. R. Walker, R. J. Vogl, and C. P. Thompson, "Autobiographical Memory: Unpleasantness Fades Faster Than Pleasantness Over Time," *Applied Cognitive Psychology* 11 (1997): 399–413.

15. D. C. Rubin, "The Distribution of Early Childhood Memories," *Memory* 8 (2000): 265–269; N. Davis, J. Gross, and H. Hayne, "Defining the Boundary of Childhood Amnesia," *Memory*, 16 (2008): 465–474.

16. J. Usher and U. Neisser, "Childhood Amnesia and the Beginnings of Memory for Four Early Life Events," *Journal of Experimental Psychology: General* 122 (1993): 155–165; J. Gross, F. Jack, N. Davis, and H. Hayne, "Do Children Recall the Birth of a Younger Sibling? Implications for the Study of Childhood Amnesia," *Memory* 21 (2013): 336–346.

17. C. M. Morrison and M. A. Conway, "First Words and First Memories," *Cognition* 116 (2010): 23–32.

18. Q. Wang, "Earliest Recollections of Self and Others in European American and Taiwanese Young Adults," *Psychological Science* 17 (2006): 708–714.

19. D. P. McAdams, "The Psychological Self as Actor, Agent, and Author," *Perspectives on Psychological Science* 8 (2013): 272–295.

20. M. A. Conway, Q. Wang, K. Hanyu, and S. Haque, "A Cross-Cultural Investigation of Autobiographical Memory," *Journal of Cross-Cultural Psychology* 36 (2005): 739–749.

21. D. C. Rubin and D. Berntsen, "Life Scripts Help to Maintain Autobiographical Memories of Highly Positive, but Not Highly Negative Events," *Memory and Cognition* 31 (2003): 1–14.

22. S. M. J. Janssen, D. C. Rubin, and P. L. St. Jacques, "The Temporal Distribution of Autobiographical Memory: Changes in Reliving and Vividness Do Not Explain the Reminiscence Bump," *Memory and Cognition* 39 (2011): 1–11; C. J. Rathbone, C. J. A. Moulin, and M. A. Conway, "Self-Centered Memories: The Reminiscence Bump and the Self," *Memory and Cognition* 36 (2008): 1403–1414.

23. S. Szkotak, "*Titanic* Survivor Recalls the Sinking 70 Years Ago," *Middletown Press*, April 14, 1982.

24. F. W. Colegrove, "The Day They Heard about Lincoln," *American Journal of Psychology* 19 (1899): 228–255.

25. R. Brown and J. Kulik, "Flashbulb Memories," *Cognition* 5 (1977): 73–99; M. McCloskey, C. G. Wible, and N. J. Cohen, "Is There a Special Flashbulb-Memory Mechanism?" *Journal of Experimental Psychology: General* 117 (1988): 171–181; J. M. Talarico and D. C. Rubin, "Confidence, Not Consistency, Characterizes Flashbulb Memories," *Psychological Science* 14 (2003): 455–461.

26. D. L. Greenberg, "President Bush's False 'Flashbulb' Memory of 9/11/01," *Applied Cognitive Psychology* 18 (2004): 363–370.

27. E. Winograd, E. Bergman, C. Schreiber, S. Palmer, and M. Weldon, "Remembering the Earthquake: Direct Experience vs. Hearing the News," *Memory* 4 (1996): 337–357; I. Nachson and I. Slavutskay-Tsukerman, "Effect of Personal Involvement in Traumatic Events on Memory: The Case of the Dolphinarium Explosion," *Memory* 18 (2010): 241–251.

28. G. Nigro and U. Neisser, "Point of View in Personal Memories," *Cognitive Psychology* 15 (1983): 467–482.

29. U. Neisser, "Snapshots or Benchmarks," In *Memory Observed,* 2nd ed., ed. U. Neisser and I. E. Hyman, Jr. (New York: Worth, 2010).

30. R. Pausch, *The Last Lecture* (New York: Hyperion, 2008).

6 When Troubling Memories Persist

1. W. James, *The Principles of Psychology* (New York: Dover, 1890).

2. I. Sotgiu and C. Mormont, "Similarities and Differences between Traumatic and Emotional Memories: Review and Directions for Future Research," *Journal of Psychology* 142 (2008): 449–469.

3. R. J. McNally, *Remembering Trauma* (Cambridge, MA: Harvard University Press, 2003).

4. E. A. Kensinger and D. L. Schacter, "Memory and Emotion," In *The Handbook of Emotion,* 3rd ed., ed. M. Lewis, J. M. Haviland-Jones, and L. F. Barrett (New York: Guilford, 2008); D. Reisberg and P. Hertel, *Memory and Emotion* (New York: Oxford University Press, 2005).

5. L. Cahill, and J. L. McGaugh, "A Novel Demonstration of Enhanced Memory Associated with Emotional Arousal," *Consciousness and Cognition* 4 (1995): 410–421.

6. J. A. Easterbrook, "The Effect of Emotion on Cue Utilization and the Organization of Behavior," *Psychological Review* 66 (1959): 183–201; S.-Å. Christianson and E. F. Loftus, "Remembering Emotional Events: The Fate of Detailed Information," *Cognition & Emotion* 5 (1991): 81–108.

7. E. F. Loftus, G. R. Loftus, and J. Messo, "Some Facts about 'Weapon Focus,'" *Law and Human Behavior* 11 (1987): 55–62.

8. E. A. Kensinger, "Remembering the Details: Effects of Emotion," *Emotion Review* 12 (2009): 99–113.

9. J. L. McGaugh, "The Amygdala Modulates the Consolidation of Memories of Emotionally Arousing Experiences," *Annual Review of Neuroscience* 27 (2004): 1–28; S. J. Lupien, F. Maheu, M. Tu, A. Fiocco, and T. E. Schramek, "The Effects of Stress and Stress Hormones on Human Cognition: Implications for the Field of Brain and Cognition," *Brain and Cognition* 65 (2007): 209–237.

10. T. W. Buchanan, "Retrieval of Emotional Memories," *Psychological Bulletin* 133 (2007): 761–779.

11. Kensinger, "Remembering the Details"; T. Sharot, M. R. Delgado, and E. A. Phelps, "How Emotion Enhances the Feeling of Remembering," *Nature Neuroscience* 7 (2004): 1376–1380.

12. M. E. McDevitt-Murphy, G. R. Parra, M. T. Shea, S. Yen, C. M. Grilo, C. A. Sanislow, et al., "Trajectories of PTSD and Substance Use Disorders in a Longitudinal Study of Personality Disorders," *Psychological Trauma: Theory, Research, and Policy* 1 (2009): 269–281; National Center for PTSD, *PTSD and Problems of Alcohol Use* (Washington, DC: Department of Veterans Affairs, 2013), http://www.ptsd.va.gov.

13. National Institute of Mental Health, *What Is Post-Traumatic Stress Disorder (PTSD)?* (Bethesda, MD: National Institute of Mental Health, 2014), http://www.nimh.nih.gov.

14. M. Spoont, P. Arbisi, S. Fu, N. Greer, S. Kehle-Forbes, L. Meis, and I. Rutks, *Screening for Post-Traumatic Stress Disorder in Primary Care: A Systematic Review* (Washington, DC: Department of Veterans Affairs, 2013); National Center for PTSD, *How Common Is PTSD?* (Washington, DC: Department of Veterans Affairs, 2013), http://www.ptsd.va.gov.

15. J. Pizarro, R. C. Silver, and J. Prause, "Physical and Mental Health Costs of Traumatic War Experiences among Civil War Veterans," *Archives of General Psychiatry* 63 (2006): 193–199.

16. R. J. McNally, N. B. Lasko, M. L. Macklin, and R. K. Pitman, "Autobiographical Memory Disturbance in Combat-Related Posttraumatic Stress Disorder," *Behavior Research and Therapy* 33 (2007): 619–630.

17. B. P. Dohrenwend, T. J. Yager, M. M. Wall, and B. G. Adams, "The Roles of Combat Exposure, Personal Vulnerability, and Involvement in Harm to Civilians or Prisoners in Vietnam War-Related Posttraumatic Stress Disorder," *Clinical Psychological Science* 1 (2013): 223–238.

18. B. T. Litz, N. Stein, E. Delaney, L. Lebowitz, and W. P. Nash, "Moral Injury and Moral Repair in War Veterans: A Preliminary Model and Intervention Strategy," *Clinical Psychology Review* 293 (2009): 695–706.

19. R. Kovic, *Born on the Fourth of July* (New York: McGraw-Hill and Akashic Press, 1976, 2005).

20. V. Hughes, "The Roots of Resilience," *Nature* 490 (2012): 165–167.

21. A. G. Harvey and R. A. Bryant, "The Relationship between Acute Stress Disorder and Posttraumatic Stress Disorder: A Prospective Evaluation of Motor Vehicle Accident Survivors," *Journal of Consulting and Clinical Psychology* 66 (1998): 507–512.

22. D. Brown, A. W. Scheflin, and D. C. Hammond, *Memory, Trauma Treatment, and the Law* (New York: W. W. Norton, 1998).

23. E. F. Loftus and K. Ketcham, *The Myth of Repressed Memory: False Memories and Allegations of Sexual Abuse* (New York: St. Martin's Press, 1994); J. Kihlstrom, "Trauma and Memory Revisited," in *Memory and Emotions: Interdisciplinary Perspectives*, ed. B. Uttl, N. Ohta, and A. L. Siegenthaler (New York: Blackwell, 2006).

24. A. Winter, *Memory: Fragments of a Modern History* (Chicago: University of Chicago Press, 2012).

25. S. Peters, G. Wyatt, and D. Finkelhor, "Prevalence," in *A Sourcebook on Child Sexual Abuse*, ed. D. Finkelhor (Beverly Hills, CA: Sage, 1986); D. Finkelhor, "The International Epidemiology of Childhood Sexual Abuse," *Child Abuse and Neglect* 18 (1994): 409–417.

26. American Psychological Association, *Understanding Child Sexual Abuse: Education, Prevention, and Recovery* (Washington, DC: American Psychological Association, 2005), http://www.apa.org.

27. American Psychological Association, *Questions and Answers about Memories of Childhood Abuse* (Washington, DC: American Psychological Association, 1995), http://www.apa.org.

28. C. J. Brainerd and V. F. Reyna, *The Science of False Memory* (New York: Oxford University Press, 2005).

29. J. W. Schooler and E. Eich, "Memory for Emotional Events," in *The Oxford Handbook of Memory*, ed. E. Tulving and F. I. M. Craik (New York: Oxford University Press, 2000).

30. L. M. Williams, "Recall of Childhood Trauma: A Prospective Study of Women's Memories of Childhood Sexual Abuse," *Journal of Consulting and Clinical Psychology* 62 (1994): 1167–1176.

31. G. S. Goodman, S. Ghetti, J. A. Quas, R. S. Edelstein, K. W. Alexander, A. D. Redlich, et al., "A Prospective Study of Memory for Childhood Sexual Abuse: New Findings Relevant to the Repressed-Memory Controversy," *Psychological Science* 14 (2003): 113–118.

32. K. A. Peace, S. Porter, and L. T. Brinke, "Are Memories for Sexually Traumatic Events 'special'? A Within-Subjects Investigation of Trauma and Memory in a Clinical Sample," *Memory* 16 (2008): 10–21.

33. W. A. Wagenaar and J. Groeneweg, "The Memory of Concentration Camp Survivors," *Applied Cognitive Psychology* 4 (1990): 77–87.

34. J. W. Schooler, M. Bendiksen, and Z. Ambadar, "Taking the Middle Line: Can We Accommodate Both Fabricated and Recovered Memories of Sexual Abuse?" in *False and Recovered Memories*, ed. M. A. Conway (Oxford: Oxford University Press, 1997).

35. J. W. Schooler, "Discovered Memories and the 'Delayed Discovery' Doctrine: A Cognitive Case-Based Analysis," in *Recovered Memories of Child Sexual Abuse*, ed. S. Taub (Springfield, IL: Charles C. Thomas, 1999).

36. I. E. Hyman and J. Pentland, "The Role of Mental Imagery in the Creation of False Childhood Memories," *Journal of Memory and Language* 35 (1996): 101–117.

37. E. F. Loftus, "The Reality of Repressed Memories," *American Psychologist* 48 (1993): 518–537; S. Porter, J. C. Yuille, and D. R. Lehman, "The Nature of Real, Implanted, and Fabricated Memories for Emotional Childhood Events: Implications for the Recovered Memory Debate," *Law and Human Behavior* 23 (1999): 517–537.

38. K. A. Wade, M. Garry, J. D. Read, and D. S. Lindsay, "A Picture Is Worth a Thousand Lies: Using False Photographs to Create False Childhood Memories," *Psychonomic Bulletin and Review* 9 (2002): 597–603.

39. M. K. Johnson, S. Hashtroudi, and D. S. Lindsay, "Source Monitoring," *Psychological Bulletin* 114 (1993): 3–28.

40. D. A. Poole, D. S. Lindsay, A. Memon, and R. Bull, "Psychotherapy and the Recovery of Memories of Childhood Sexual Abuse: US and British Practitioners' Opinions, Practices, and Experiences," *Journal of Consulting and Clinical Psychology* 63 (1995): 426–437.

41. E. F. Loftus, "Make-Believe Memories," *American Psychologist* 58 (2003): 867–873.

42. S. A. Clancy, *Abducted: How People Come to Believe They Were Kidnapped by Aliens* (Cambridge, MA: Harvard University Press, 2005); Loftus and Ketcham, *The Myth of Repressed Memory*.

43. S. J. Lynn, T. G. Lock, B. Myers, and D. G. Payne, "Recalling the Unrecallable: Should Hypnosis Be Used to Recover Memories in Psychotherapy?" *Current Directions in Psychological Science* 6 (1997): 79–83.

44. J. F. Kihlstrom, "Hypnosis, Memory, and Amnesia," *Philosophical Transactions of the Royal Society: Biological Sciences* 372 (1997): 1727–1732.

45. S. O. Lilienfeld, "Psychological Treatments That Cause Harm," *Perspectives on Psychological Science* 2 (2007): 53–70.

46. Winter, *Memory*.

47. E. Geraerts, J. W. Schooler, H. Merckelbach, M. Jelicic, B. J. A. Hauer, and Z. Ambadar, "The Reality of Recovered Memories: Corroborating Continuous and Discontinuous Memories of Childhood Sexual Abuse," *Psychological Science* 18 (2007): 564–568.

48. R. J. McNally and E. Geraerts, "A New Solution to the Recovered Memory Debate," *Perspectives on Psychological Science* 4 (2009): 126–134.

49. Schooler, et al., "Taking the Middle Line."

50. H. L. Roediger III and D. A. Gallo, "False Memory," in *Encyclopedia of Psychology*, ed. A. G. Kazdin (New York: Oxford University Press, 2000).

51. American Psychological Association, *Questions and Answers about Memories of Childhood Abuse*.

52. L. Patihis, L. Y. Ho, I. W. Tingen, S. O. Lilienfeld, and E. F. Loftus, "Are the 'Memory Wars' Over? A Scientist-Practitioner Gap in Beliefs about Repressed Memory," *Psychological Science* 25 (2014): 519–530.

53. M. L. Howe, "The Adaptive Nature of Memory and Its Illusions," *Current Directions in Psychological Science* 20 (2011): 312–315; M. K. Johnson, "Memory and Reality," *American Psychologist* 61 (2006): 760–771.

54. D. L. Schacter, S. A. Guerin, and P. L. St. Jacques, "Memory Distortion: An Adaptive Perspective," *Trends in Cognitive Science* 15 (2011): 467–474.

55. S. Joslyn, L. Carlin, and E. F. Loftus, "Remembering and Forgetting Childhood Sexual Abuse," *Memory* 5 (1997): 703–724; McNally, *Remembering Trauma*.

56. T. Dalgleish, B. Hauer, and W. Kuyken, "The Mental Regulation of Autobiographical Recollection in the Aftermath of Trauma," *Current Direction in Psychological Science* 17 (2008): 259–263.

57. P. A. Ornstein, S. J. Ceci, and E. F. Loftus, "Comment on Albert, Brown, and Courtois (1998): The Science of Memory and the Practice of Psychotherapy," *Psychology, Public Policy, and Law* 4 (1998): 996–1010.

58. E. Tulving and D. M. Thompson, "Encoding Specificity and Retrieval Processes in Episodic Memory," *Psychological Review* 80 (1973): 352–373.

59. Schooler and Eich, "Memory for Emotional Events."

60. R. K. Pitman, K. M. Sanders, R. M. Zusman, A. R. Healy, F. Cheema, N. B. Lasko, et al., "Pilot Study of Secondary Prevention of Posttraumatic Stress Disorder with Propranolol," *Biological Psychiatry* 51 (2002): 189–192.

61. R. K. Pitman, "Will Reconsolidation Blockade Offer a Novel Treatment for Posttraumatic Stress Disorder?" *Frontiers in Behavioral Neuroscience* 5 (2011), doi: 10.3389/fnbeh.2011.00011.

62. D. Clark, C. G. Fairburn, and M. G. Gelder, eds., *Science and Practice of Cognitive Behavior Therapy* (Oxford: Oxford University Press, 1997).

7 Understanding the Reality of Amnesia

1. R. C. Cantu, "Second-Impact Syndrome," *Clinics in Sports Medicine* 17 (1998): 37–44; M. W. Kirkwood, K. O. Yeates, and P. E. Wilson, "Pediatric Sport-Related Concussion: A Review of the Clinical Management of an Oft-Neglected Population," *Pediatrics* 117 (2006): 1359–1371.

2. S. Baxendale, "Memories Aren't Made of This: Amnesia at the Movies," *British Medical Journal* 329 (2004): 1480–1483.

3. B. Milner, "Amnesia Following Operation on the Temporal Lobes," in *Amnesia*, ed. C. W. M. Whitty and O. L. Zangwill (London: Butterworth, 1966); A. D. Baddeley and E. K. Warrington, "Amnesia and the Distinction between Long- and Short-Term Memory," *Journal of Verbal Learning and Verbal Behavior* 9 (1970): 176–189.

4. J. F. Kihlstrom, "Dissociative Disorders," *Annual Review of Clinical Psychology* 1 (2005): 227–253.

5. J. M. Murre and D. P. Sturdy, "The Connectivity of the Brain: Multi-level Quantitative Analysis," *Biological Cybernetics* 73 (1995): 529–545; S. Herculano-Houzel, "The Human Brain in Numbers: A Linearly Scaled-

Up Primate Brain," *Frontiers in Human Neuroscience* 31 (2009), doi: 10.3389/neuro.09.031.2009.

6. A. Stracciari, E. Ghidoni, M. Guarino, M. Poletti, and P. Pazzaglia, "Post-Traumatic Retrograde Amnesia with Selective Impairment of Autobiographical Memory," *Cortex* 30 (1994): 459–468.

7. A. Burns and M. Burns, *Surviving Amnesia: Mind Over Memory* (Adam Burns, Amazon Digital Services, 2012).

8. A. J. Parkin, *Memory and Amnesia: An Introduction* (Oxford: Blackwell, 1987); Brain Injury Association of America, *About Brain Injury* (2013), http://www.biause.org.

9. T.-A. Ribot, *Diseases of Memory* (New York: Appleton, 1882).

10. J. Barbizet, *Human Memory and Its Pathology* (San Francisco: Freeman, 1970).

11. A. S. Brown, "Consolidation Theory and Retrograde Amnesia in Humans," *Psychonomic Bulletin and Review* 9 (2002): 403–425.

12. S. Finger and F. Zaromb, "Benjamin Franklin and Shock-Induced Amnesia," *American Psychologist* 61 (2006): 240–248.

13. M. O'Conner et al., "A Dissociation between Anterograde and Retrograde Amnesia after Treatment with Electroconvulsive Therapy: A Naturalistic Investigation," *Journal of ECT* 24 (2008): 146–151.

14. S. Zola-Morgan, L. R. Squire, and D. G. Amarai, "Human Amnesia and the Medial Temporal Region: Enduring Memory Impairment Following a Bilateral Lesion Limited to Field CA1 of the Hippocampus." *Journal of Neuroscience* 6 (1986): 2950–2967; L. R. Squire and S. Zola-Morgan, "The Medial Temporal Memory System," *Science* 253 (1991): 1380–1386.

15. A. S. Brown, "Transient Global Amnesia," *Psychonomic Bulletin and Review* 5 (1998): 401–427.

16. M. Park, "Sex, Then Amnesia … and It's No Soap Opera," CNN.com, Nov. 5, 2009.

17. J. R. Hodges and C. D. Ward, "Observations during Transient Global Amnesia: A Behavioral and Neuropsychological Study of Five Cases," *Brain* 112 (1989): 595–620.

18. O. Sacks, *An Anthropologist on Mars: Seven Paradoxical Tales* (New York: Alfred A. Knopf, 1995).

19. D. Wearing, *Forever Today: A Memoir of Love and Amnesia* (London: Doubleday, 2005).

20. B. A. Wilson and D. Wearing, "Prisoner of Consciousness: A State of Just Awakening Following Herpes Simplex Encephalitis," in *Broken Memories: Case Studies in Memory Impairment*, ed. R. Campbell and M. A. Conway (Cambridge, MA: Blackwell, 1995).

21. O. Sacks, *The Man Who Mistook His Wife for a Hat and Other Clinical Tales* (New York: Harper & Row, 1970).

22. Parkin, *Memory and Amnesia*.

23. R. J. McNally, *Remembering Trauma* (Cambridge, MA: Harvard University Press, 2003).

24. M. D. Kopelman, "The Assessment of Psychogenic Amnesia," in *Handbook of Memory Disorders*, ed. A. D. Baddeley, B. A. Wilson, and F. N. Watts (Chichester: John Wiley & Sons, 1995).

25. Kihlstrom, "Dissociative Disorders."

26. J. F. Kihlstron and D. L. Schacter, "Functional Disorders of Autobiographical Memory," in *Handbook of Memory Disorders*, ed. A. D. Baddeley, B. A. Wilson, and F. N. Watts (Chichester: John Wiley & Sons, 1995); M. Kritchevsky, J. Chang, and L. R. Squire, "Functional Amnesia: Clinical Description and Neuropsychological Profile of 10 Cases," *Learning and Memory* 11 (2011): 213–226.

27. J. F. Kihlstrom, M. L. Glisky, and M. J. Angiulo, "Dissociative Tendencies and Dissociative Disorders," *Journal of Abnormal Psychology* 103 (1994): 117–124.

28. Kihlstron and Schacter, "Functional Disorders of Autobiographical Memory."

29. R. R. Grinker and J. P. Spiegel, *Men under Stress* (New York: McGraw-Hill, 1945).

30. D. L. Schacter, P. L. Wang, E. Tulving, and M. Freedman, "Functional Retrograde Amnesia: A Quantitative Case Study," *Neuropsychologia* 20 (1982): 523–532.

31. "Dissociative Disorders," in *Diagnostic and Statistical Manual of Mental Disorders*, 5th ed. (Washington, DC: American Psychiatric Publishing, 2013).

32. D. L. Schacter, J. F. Kihlstrom, L. C. Kihlstrom, and M. B. Berren, "Autobiographical Memory in a Case of Multiple Personality Disorder," *Journal of Abnormal Psychology* 98 (1989): 508–514.

33. M. J. Nissen, J. L. Ross, D. B. Willingham, T. B. Mackenzie, and D. L. Schacter, "Memory and Awareness in a Patient with Multiple Personality Disorder," *Brain and Cognition* 8 (1988): 117–134; E. Eich, D. Macaulay, R. J. Loewenstein, and P. H. Dihle, "Memory, Amnesia, and Dissociative Identity Disorder," *Psychological Science* 8 (1997): 417–422.

34. D. L. Schacter, "Amnesia and Crime: How Much Do We Really Know?" *American Psychologist* 41 (1986): 286–295.

35. S. Corkin, *Permanent Present Tense: The Unforgettable Life of the Amnesic Patient, H.M.* (New York: Basic Books, 2013).

36. Ibid.

37. Ibid.

38. B. Milner, "Visually-Guided Maze Learning in Man: Effects of Bilateral Hippocampal, Bilateral Frontal, and Unilateral Cerebral Lesions," *Neuropsychologia* 3 (1965): 317–338; S. Corkin, "Tactually-Guided Maze Learning in Man: Effects of Unilateral Cortical Excisions and Bilateral Hippocampal Lesions," *Neuropsychologia* 3 (1965): 339–351.

39. J. D. E. Gabrieli et al., "Dissociations among Structural-Perceptual, Lexical-Semantic, and Event-Fact Memory Systems in Alzheimer, Amnesic, and Normal Subjects," *Cortex* 30 (1994): 75–103.

40. G. O'Kane, E. A. Kensinger, and S. Corkin, "Evidence for Semantic Learning in Profound Amnesia: An Investigation with Patient H.M.," *Hippocampus* 14 (2004): 417–425.

41. Corkin, *Permanent Present Tense.*

42. P. J. Hilts, *Memory's Ghost: The Strange Tale of Mr. M. and the Nature of Memory* (New York: Simon & Schuster, 1995).

43. A. Nocenti, "Christopher Nolan's Revenge Redux," *Independent,* March 2001, 32–35.

44. M. Baur, "We All Need Mirrors to Remind Us Who We Are: Inherited Meaning and Inherited Selves in *Memento,*" in *Movies and the Meaning of Life,* ed. K. A. Blessing and P. J. Tudico (Chicago: Open Court Press, 2005).

45. S. Meck, *I Forgot to Remember: A Memoir of Amnesia* (New York: Simon & Schuster, 2014).

46. Ibid.

47. B. A. Wilson, *Memory Rehabilitation: Integrating Theory and Practice* (New York: Guilford Press, 2009).

8 Senior Moments, Forgetfulness, and Dementia

1. R. J. Blendon, J. M. Benson, E. M. Wikler, K. J. Weldon, J. Georges, M. Baumgart, and B. A. Kallmyer, "The Impact of Experience with a Family Member with Alzheimer's Disease on Views about the Disease across Five Countries," *International Journal of Alzheimer's Disease* (2012), doi: 10.1155/2012/903645; S. Magnussen, J. Andersson, C. Cornoldi, R. De Beni, T. Endestad, G. S. Goodman, et al., "What People Believe about Memory," *Memory* 14 (2006): 595–613.

2. P. V. Rabins, *Memory: The Johns Hopkins White Papers* (Baltimore, MD: Johns Hopkins Medicine, 2014).

3. T. Friedman, *You Are Not Alone ... 1,000 Unforgettable Senior Moments* (New York: Workman Publishing, 2006).

4. G. O. Einstein and M. A. McDaniel, *Memory Fitness: A Guide for Successful Aging* (New Haven, CT: Yale University Press, 2004).

5. L. L. Light, "Memory and Aging," in *Memory*, ed. E. L. Bjork and R. A. Bjork (San Diego, CA: Academic Press, 1996); T. Salthouse, *Theoretical Perspectives on Cognitive Aging* (Hillsdale, NJ: Erlbaum, 1991).

6. M. Rönnlund, L. Nyberg, L. Bäckman, and L.-G. Nilsson, "Stability, Growth, and Decline in Adult Life-span Development of Declarity of Memory: Cross Sectional and Longitudinal Data from a Population-Based Sample," *Psychology and Aging* 20 (2005): 3–18.

7. D. C. Park, R. Nisbett, and T. Hedden, "Aging, Culture, and Cognition," *Journals of Gerontology Series B: Psychological Sciences and Social Sciences* 54B (1999): 75–84.

8. Rabins, *Memory*.

9. T. Hedden and J. D. E. Gabrieli, "Insights into the Aging Mind: A View from Cognitive Neuroscience," *Nature Reviews Neuroscience* 5 (2004): 87–96; Einstein and McDaniel, *Memory Fitness*.

10. M. Mather, T. Canli, T. English, S. Whitfield, P. Wais, K. Ochsner, et al., "Amygdala Responses to Emotionally Valenced Stimuli in Older and Younger Adults," *Psychological Science* 15 (2004): 259–263.

11. S. T. Charles, M. Mather, and L. L. Carstensen, "Aging and Emotional Memory: The Forgettable Nature of Negative Images for Older Adults," *Journal of Experimental Psychology: General* 132 (2003): 310–324.

12. Einstein and McDaniel, *Memory Fitness*.

13. D. L. Schacter, *Searching for Memory: The Brain, the Mind, and the Past* (New York: Basic Books, 2004).

14. Hedden and Gabrieli, "Insights into the Aging Mind."

15. T. A. Salthouse, "The Processing-Speed Theory of Adult Age Differences in Cognition," *Psychological Review* 103 (1996): 403–428.

16. J. McCabe and M. Hartman, "An Analysis of Age Differences in Perceptual Speed," *Memory and Cognition* 36 (2008): 1495–1508.

17. L. Hasher and R. T. Zacks, "Working Memory, Comprehension, and Aging: A Review and a New View," in *The Psychology of Learning and Motivation: Advances in Research and Theory*, ed. G. H. Bower (San Diego, CA: Academic Press, 1988).

18. B. Levy, "Improving Memory in Old Age through Implicit Self-Stereotyping," *Journal of Personality and Social Psychology* 71 (1996): 1092–1107.

19. B. Levy and E. Langer, "Aging Free from Negative Stereotypes: Successful Memory in China and among the American Deaf," *Journal of Personality and Social Psychology* 66 (1994): 989–997.

20. F. I. M. Craik, "A Functional Account of Age Differences in Memory," in *Human Memory and Cognitive Capabilities: Mechanisms and Performances*, ed. F. Klix and H. Hagendorf (New York: Elsevier Science, 1986).

21. L.-G. Nilsson, "Memory Function in Normal Aging," *Acta Neurologica Scandinavica* 107 (2003): 7–13.

22. Hedden and Gabrieli, "Insights into the Aging Mind."

23. T. Singer, P. Verhaeghen, P. Ghisletta, U. Lindenberger, and P. B. Baltes, "The Fate of Cognition in Very Old Age: Six-Year Longitudinal Findings in the Berlin Aging Study (BASE)," *Psychology and Aging* 18 (2003): 318–331; L.-G. Nilsson, L. Bäckman, K. Erngrund, L. Nyberg, R. Adolfsson, G. Bucht, et al., "The Betula Prospective Cohort Study: Memory, Health, and Aging," *Aging, Neuropsychology, and Cognition* 4 (1997): 1–32.

24. D. J. Dahlgren, "Impact of Knowledge and Age on Tip-of-the-Tongue Rates," *Experimental Aging Research* 24 (1998): 139–153.

25. A. T. Welford, "Changes in Performance with Age: An Overview," in *Aging and Human Performance*, ed. N. Charness (New York: Wiley, 1985); A. M. Brickman and Y. Stern, "Aging and Memory in Humans," *Encyclopedia of Neuroscience* 1 (2009): 175–180.

26. Alzheimer's Association "Mild Cognitive Impairment," http://www.alz.org/dementia/mild-cognitive-impairment-mci.asp; R. J. Caselli, K.

Chen, D. E. C. Locke, W. Lee, A. Roontiva, D. Bandy, et al., "Subjective Cognitive Decline: Self and Informant Comparisons," *Alzheimer's and Dementia* 10 (2014): 93–98.

27. R. C. Petersen, G. E. Smith, S. C. Waring, R. J. Ivnik, E. G. Tangalos, and E. Kokmen, "Mild Cognitive Impairment: Clinical Characterization and Outcome," *Archives of Neurology* 56 (1999): 303–308.

28. Z. S. Nasreddine, N. A. Phillips, V. Bédirian, S. Charbonneau, V. Whitehead, I. Collin, et al. "The Montreal Cognitive Assessment (MoCA): A Brief Screening Tool for Mild Cognitive Impairment," *Journal of the American Geriatrics Society* 53 (2005): 695–699.

29. E. J. Mufson, L. Binder, S. E. Counts, S. T. DeKosky, L. deToledo-Morrell, S. D. Ginsberg, et al., "Mild Cognitive Impairment: Pathology and Mechanisms," *Acta Neuropathologica* 123 (2012): 13–30; L.-O. Wahlund, E. Pihlstrand, and M. E. Jönhagen, "Mild Cognitive Impairment: Experience from a Memory Clinic," *Acta Neurologica Scandinavica* 107 (2003): 21–24.

30. J. C. Morris, M. Storandt, J. P. Miller, D. W. McKeel, J. L. Price, E. H. Rubin, and L. Berg, "Mild Cognitive Impairment Represents Early-Stage Alzheimer Disease," *Archives of Neurology* 58 (1999): 397–405.

31. V. Jelic and B. H. Winblad, "Treatment of Mild Cognitive Impairment: Rationale, Present, and Future Strategies," *Acta Neurologica Scandinavica* 107 (2003): 83–93.

32. American Psychiatric Association, *Diagnostic and Statistical Manual of Mental Disorders*, 5th ed. (Washington, DC: American Psychiatric Publishing, 2013).

33. Rabins, *Memory*.

34. Centers for Disease Control and Prevention, "Stroke" (2014), http://www.cdc.gov/stroke/.

35. National Institute of Neurological Disorders and Stroke, "NINDS Multi-Infarct Dementia Information Page" (2014), http://www.ninds.nih.gov/index.htm.

36. Ibid.

37. K. Butler, "What Broke My Father's Heart," *New York Times*, June 18, 2010; K. Butler, *Knocking on Heaven's Door* (New York: Scribner, 2013).

38. S. Foundas, "Michael Haneke on *Amour*," *Village Voice Blogs*, December 20, 2012.

39. American Psychiatric Association, *Diagnostic and Statistical Manual of Mental Disorders*, 5th ed.; Alzheimer's Association, "Mild Cognitive Impairment."

40. E. Morris, *Dutch: A Memoir of Ronald Reagan* (New York: Random House, 1999).

41. Rabins, *Memory*.

42. American Psychiatric Association, *Diagnostic and Statistical Manual of Mental Disorders*, 5th ed.

43. Alzheimer's Association, "Alzheimer's Disease Facts and Figures," *Alzheimer's and Dementia* 10 (2014): 47–92.

44. G. M. McKhann, D. S. Knopman, K. Chertkow, B. T. Hyman, C. R. Jack, Jr., C. H. Kawas, et al., "The Diagnosis of Dementia Due to Alzheimer's Disease: Recommendations from the National Institute on Aging-Alzheimer's Association Workgroups on Diagnostic Guidelines for Alzheimer's Disease *Alzheimer's and Dementia* 7 (2011): 263–269.

45. M. Storandt, "Cognitive Deficits in the Early Stages of Alzheimer's Disease," *Current Directions in Psychological Science* 17 (2008): 198–202; B. J. Small, J. L. Mobly, E. J. Laukka, S. Jones, and L. Bäckman, "Cognitive Deficits in Preclinical Alzheimer's Disease," *Acta Neurologica Scandinavica* 107 (2003): 29–33.

46. P. Garrard, L. M. Malony, J. R. Hodges, and K. Patterson, "The Effects of Very Early Alzheimer's Disease on the Characteristics of Writing by a Renowned Author," *Brain* 128 (2005): 250–260.

47. R. Sherva, Y. Tripodis, D. A. Bennett, L. B. Chibnik, P. K. Crane, P. L. de Jager, et al., "Genome-Wide Association Study of the Rate of Cognitive Decline in Alzheimer's Disease," *Alzheimer's and Dementia* 10 (2014):

45–52; American Psychiatric Association, *Diagnostic and Statistical Manual of Mental Disorders*, 5th ed.

48. Rabins, *Memory*.

49. D. R. Addis, D. C. Sacchetti, B. A. Ally, A. E. Budson, and D. L. Schacter, "Episodic Simulation of Future Events Is Impaired in Mild Alzheimer's Disease," *Neuropsychologia* 47 (2009): 2660–2671.

50. Rabins, *Memory*.

51. C. Bernard, C. Helmer, B. Dilharreguy, H. Amieva, S. Auriacombe, J.-F. Dartigues, et al., "Time Course of Brain Volume Changes in the Preclinical Phase of Alzheimer's Disease," *Alzheimer's and Dementia* 10 (2014): 141–151.

52. K. Henriksen, S. E. O'Bryant, H. Hampel, J. Q. Trojanowski, T. J. Montine, A. Jeromin, et al., "The Future of Blood-Based Biomarkers for Alzheimer's Disease," *Alzheimer's and Dementia* 10 (2014): 115–131.

53. J. C. de la Torre, "The Vascular Hypothesis of Alzheimer's Disease: Bench to Bedside to Beyond," *Neurodegenerative Diseases* 7 (2010): 116–121; D. A. Drachman, "The Amyloid Hypothesis, Time to Move On: Amyloid Is the Downstream Result, Not Cause, of Alzheimer's Disease," *Alzheimer's and Dementia* 10 (2014): 372–380.

54. I. Raiha, J. Kaprio, M. Koskenvuo, T. Rajala, and L. Sourander, "Alzheimer's Disease in Finnish Twins," *Lancet* 347 (1996): 573–578.

55. American Psychiatric Association, *Diagnostic and Statistical Manual of Mental Disorders*, 5th ed.; A. Terracciano, A. R. Sutin, Y. An, R. J. O'Brien, L. Ferrucci, A. B. Zonderman, and S. M. Resnick, "Personality and Risk of Alzheimer's Disease: New Data and Meta-Analysis," *Alzheimer's and Dementia* 10 (2014): 179–186.

56. Rabins, *Memory*.

57. G. Epstein-Lubow, "A Family Disease: Witnessing Firsthand the Toll That Dementia Takes on Caregivers," *Health Affairs* 33 (2014): 708–711.

58. L. L. Cuddy and J. Duffin, "Music, Memory, and Alzheimer's Disease: Is Music Recognition Spared in Dementia, and Can It Be Assessed?" *Medical Hypotheses* 64 (2005): 229–235.

59. A. Cowles, W. W. Beatty, S. J. Nixon, L. J. Lutz, J. Paulk, K. Paulk, and E. D. Ross, "Musical Skill in Dementia: A Violinist Presumed to Have Alzheimer's Disease Learns to Play a New Song," *Neurocase* 9 (2005): 493–503; H. A. Crystal, E. Grober, and D. Masur, "Preservation of Musical Memory in Alzheimer's Disease," *Journal of Neurology, Neurosurgery, and Psychiatry* 52 (1989): 1415–1416.

60. O. Sacks, *Musicophilia: Tales of Music and the Brain* (New York: Alfred A. Knopf, 2007).

61. J. A. Hehman, T. P. German, and S. B. Klein, "Impaired Self-Recognition from Recent Photographs in a Case of Late-Stage Alzheimer's Disease," *Social Cognition* 23 (2005): 118–123; S. B. Klein, L. Cosmides, and K. A. Costabile, "Preserved Knowledge of Self in a Case of Alzheimer's Dementia," *Social Cognition* 21 (2003): 157–165.

62. V. E. Sturm, J. S. Yokoyama, W. W. Seeley, J. H. Kramer, B. L. Miller, and K. P. Rankin, "Heightened Emotional Contagion in Mild Cognitive Impairment and Alzheimer's Disease Is Associated with Temporal Lobe Degeneration," *Proceedings of the National Academy of Sciences* 110 (2013): 9944–9949.

63. S. O'Driscoll, "Love amid Dementia: New Romances Can Comfort Patients but Unsettle Families," *Hartford Courant*, December 2, 2007.

64. H. Hatfield, "The Emotional Toll of Alzheimer's," WebMD (2008), htttp://www.webmd.com/alzheimers/features/emotional-toll-of-alzheimers; S. O'Driscoll, "Love amid Dementia."

65. D. Bair, *Calling It Quits: Late Life Divorce and Starting Over* (New York: Random House, 2007).

66. T. A. Salthouse, "Mental Exercise and Mental Aging," *Perspectives on Psychological Science* 1 (2006): 68–87; M. A. McDaniel and J. M. Bugg, "Memory Training Interventions: What Has Been Forgotten?" *Journal of Applied Research in Memory and Cognition* 1 (2012): 45–50; T. S. Redick, Z.

Shipstead, T. L. Harrison, K. L. Hicks, D. E. Fried, D. Z. Hambrick, et al., "No Evidence of Intelligence Improvement after Working Memory Training: A Randomized, Placebo-Controlled Study," *Journal of Experimental Psychology: General* 142 (2013): 359–379.

67. Einstein and McDaniel, *Memory Fitness.*

68. J. C. Smith, K. A. Nielson, J. L. Woodard, M. Seidenberg, S. Durgerian, et al., "Physical Activity Reduced Hippocampal Atrophy in Elders at Genetic Risk for Alzheimer's Disease," *Frontiers in Aging Neuroscience* 6 (2014): 1–7, doi: 10.3389/fnagi.2014.00061.

69. C. Schooler, "Use It—and Keep It, Longer, Probably: A Reply to Salthouse (2006)," *Perspectives on Psychological Science* 2 (2007): 24–29.

70. D. A. Snowdon, "Healthy Aging and Dementia: Findings from the Nun Study," *Annals of Internal Medicine* 139 (2003): 450–454.

71. Y. Stern, "Cognitive Reserve," *Neuropsychologia* 47 (2009): 2015–2028.

72. E. J. Rogalski, T. Gefen, J. Shi, M. Samimi, E. Bigio, S. Weintraub, et al., "Youthful Memory Capacity in Old Brains: Anatomic and Genetic Clues from the Northwestern SuperAging Project," *Journal of Cognitive Neuroscience* 25 (2012): 29–36.

73. C. Schooler, M. S. Mulatu, and G. Oates, "The Continuing Effects of Substantively Complex Work on the Intellectual Functioning of Older Workers," *Psychology and Aging* 14 (1999): 483–506.

74. D. C. Park, J. Lodi-Smith, L. Drew, S. Haber, A. Hebrank, G. N. Bischof, and W. Aamodt, "The Impact of Sustained Engagement on Cognitive Function in Older Adults: The Synapse Project," *Psychological Science* 25 (2014): 103–112.

75. C. Hertzog, A. F. Kramer, R. S. Wilson, and U. Lindenberger, "Enrichment Effects on Adult Cognitive Development," *Psychological Science in the Public Interest* 9 (2009): 1–65.

76. Hedden and Gabrieli, "Insights into the Aging Mind."

77. O. Sacks, "The Joy of Old Age (No Kidding)," *New York Times*, Sunday Review/The Opinion Pages, July 6, 2013.

Epilogue

1. M. C. Carnes, ed., *Past Imperfect: History According to the Movies* (New York: Henry Holt, 1995); R. B. Toplin, *History in Hollywood*, 2nd ed. (Urbana: University of Illinois Press, 2010).

2. A. C. Butler, F. M. Zaromb, K. B. Lyle, and H. L. Roediger, "Using Popular Films to Enhance Classroom Learning," *Psychological Science* 20 (2009): 1161–1168; S. Umanath, A. C. Butler, and E. J. Marsh, "Positive and Negative Effects of Monitoring Popular Films for Historical Inaccuracies," *Applied Cognitive Psychology* 26 (2012): 556–567.

3. Christopher Nolan, quoted in G. Lewis-Kraus, "The Conjurer," *New York Times Magazine*, November 2, 2014.

Index

Printed in the United States
by Baker & Taylor Publisher Services